UNIVERSITY OF STRATHCLYDE

Report 182

London, 1998

Pumping stations

Design for improved buildability
and maintenance

S T Wharton MSc CEng FICE FCIWEM
P Martin BSc CEng MICE MCIWEM
T J Watson BSc CEng MICE FCIWEM

CIRIA sharing knowledge ■ building best practice

6 Storey's Gate, Westminster, London SW1P 3AU
TELEPHONE 0171 222 8891   FAX 0171 222 1708
EMAIL switchboard@ciria.org.uk
WEBSITE www.ciria.org.uk

Summary

Recommendations are given for improved buildability and maintenance of sewage pumping stations. The intention is to make the designer aware of particular operational and construction features, and to provide guidance on how to avoid potential problems.

The topics covered include design for ease, speed and economy of construction and plant installation; design for ease, economy and reliability of routine operation and maintenance; design for ease and convenience of other maintenance; design for ease, speed and economy of rehabilitation; and balancing whole-life costs. Checklists and design considerations for sewage pumping stations are included.

Pumping stations – design for improved buildability and maintenance
Construction Industry Research and Information Association
Report 182

© CIRIA 1998
ISBN 0-86017-495-6

Construction Industry Research and Information Association
6 Storey's Gate, Westminster, London SW1P 3AU
Telephone: 0171 222 8891   Facsimile: 0171 222 1708
Email: switchboard@ciria.org.uk

Published by CIRIA. All rights reserved. No part of this publication may be reproduced or transmitted in any form or by any means, including photocopying and recording, without the written permission of the copyright holder, application for which should be addressed to the publisher. Such written permission must also be obtained before any part of this publication is stored in a retrieval system of any nature.

Keywords
Pumping stations, pumping, sewage, design, buildability, construction, rehabilitation, repairs.

Reader interest	Classification	
Design, specification, construction, supervising and operations engineers involved with sewage pumping stations.	AVAILABILITY	Unrestricted
	CONTENT	Best practice guidance
	STATUS	Committee guided
	USER	Construction professionals

Foreword

The project leading to this book is part of a CIRIA programme, the aim of which is to improve the "buildability" of structures by providing feedback from construction sites on identified construction difficulties, which can be eased or eliminated if they are recognised at the feasibility and detail design stages. Of particular relevance are CIRIA's projects on the application of standardisation, pre-assembly and modularisation to provide better-value construction projects.

This book is the result of a research project carried out under contract to CIRIA by Binnie Black & Veatch (formerly Binnie & Partners). The book's purpose is to provide guidance, with background information, to designers for improving the ease of construction and maintenance of sewage pumping stations.

A review of relevant published information from the UK and overseas was carried out to identify good practice. The starting point for this was *Practical buildability*[3] and *Design of low-lift pumping stations with particular application to wastewater*[4], together with *A guide to sewerage operational practices*[6] and *BS 8005, Part 2, 1987, Sewerage. Part 2: Guide to pumping stations and pumping mains*[5].

Contacts within the Steering Group member organisations provided data and case histories covering pumping station design, construction, operation, maintenance and rehabilitation. Additional contacts with pump manufacturers and other specialists were also made. A questionnaire was prepared to assist the people contacted to compile the required data, and to form the basis for discussion during interview visits. During these visits other relevant data and drawings were obtained. Conclusions drawn from this data form the basis of this book.

Following CIRIA's usual practice, the research project was guided by a Steering Group, which comprised:

J P Cowan	consulting engineer
W Crawford	Yorkshire Water Services Ltd
P Dickens	LTG Consultants
P S Durrant	Rofe Kennard & Lapworth
C Kemp	Central Regional Council
D R Lamont	Health & Safety Executive
P Murphy	Laing Civil Engineering
R Pepper	Mono Pumps (formerly H2O Waste-Tec)
K Preston	Northumbrian Water Ltd
M Roche	Bechtel Water Technology Ltd (formerly North West Water Engineering Ltd)
M R Sinclair	ITT Flygt Ltd
C R S Tordoff	Mott MacDonald Group
C J Waters	Thames Water Utilities Ltd
S E Wielebski	Wimpey Homes

CIRIA's research manager for the project was Dr B W Staynes.

The project was funded by Bechtel Water Technology Ltd, ITT Flygt Ltd, Mono Pumps, Northumbrian Water Ltd, SADWAS, Thames Water Utilities Ltd and Yorkshire Water Services Ltd.

Contents

SUMMARY 2

FOREWORD 3

FIGURES 8
TABLES 8

GLOSSARY 9

ABBREVIATIONS 11

1 INTRODUCTION 12
    1.1 DEFINITIONS 12
    1.2 BACKGROUND TO THE STUDY 13
    1.3 FEEDBACK 14
    1.4 SCOPE OF THE STUDY 14
    1.5 STRUCTURE OF THE BOOK 15

2 CONSTRUCTION AND PLANT INSTALLATION 17
    2.1 INTRODUCTION 17
    2.2 GENERAL DESIGN CONSIDERATIONS 17
    2.3 HEALTH, SAFETY AND WELFARE DESIGN 20
    2.4 CONSTRUCTION METHODS 20
    2.5 PLANT SELECTION 23
    2.6 SUPERSTRUCTURE DESIGN 29
    2.7 WET WELL DESIGN 29
    2.8 SERVICES 31
    2.9 COMMUNICATION 31
    2.10 CONSTRUCTION (DESIGN AND MANAGEMENT) REGULATIONS[1] 32
    2.11 CHECKLIST 34

3 ROUTINE OPERATION AND MAINTENANCE 37
    3.1 INTRODUCTION 37
    3.2 RECOMMENDED MAINTENANCE VISITS AND ACTIVITIES 37
    3.3 ACTUAL MAINTENANCE VISITS AND ACTIVITIES 37
    3.4 SAFETY AND RELIABILITY 40
    3.5 COMMON PROBLEMS 41
    3.6 SUMP DESIGN 42
    3.7 PLANT SELECTION 43
    3.8 PUMPING MAIN DESIGN 43
    3.9 STATION PIPEWORK 45
    3.10 PROVISION OF ACCESS FOR ROUTINE MAINTENANCE OPERATIONS 45
    3.11 ELECTRICAL INSTALLATION 46
    3.12 AUTOMATION 47
    3.13 INSTRUMENTATION, CONTROL AND AUTOMATION (ICA) 48
    3.14 HAZARDOUS AREAS 51
    3.15 COMMUNICATION 51
    3.16 CASE HISTORIES 52
    3.17 CHECKLIST 53

4 MAJOR REPAIRS AND MAINTENANCE ... 56

4.1 INTRODUCTION ... 56
4.2 COMMON PROBLEMS ... 56
4.3 ACCESS, WORKING SPACE AND VENTILATION ... 56
4.4 COMMUNICATION ... 57
4.5 MAINTAINING FLOWS DURING MAINTENANCE OPERATIONS ... 57
4.6 ELECTRICAL AND CONTROL SYSTEMS ... 58
4.7 MAINTENANCE CONTRACTS ... 58
4.8 SIMPLICITY AND STANDARDISATION ... 59
4.9 CASE HISTORIES ... 60
4.10 CHECKLIST ... 62

5 REHABILITATION ... 64

5.1 INTRODUCTION ... 64
5.2 REPLACEMENT OF WORN-OUT PLANT ... 64
5.3 UPGRADING EXISTING PLANT ... 64
5.4 UTILISATION OF NEW TECHNOLOGY ... 65
5.5 IMPROVED EFFICIENCY AND REDUCED OPERATION AND MAINTENANCE COSTS ... 65
5.6 DESIGN OPTIONS ... 65
5.7 CURRENT TRENDS ... 66
5.8 BELOW-GROUND STRUCTURE ... 66
5.9 LAND ACQUISITION ... 67
5.10 SUPERSTRUCTURE ... 67
5.11 LIFTING EQUIPMENT ... 68
5.12 PUMPING MAINS ... 68
5.13 CASE HISTORIES ... 68
5.14 CHECKLIST ... 69

6 BALANCING WHOLE-LIFE COSTS ... 70

6.1 INTRODUCTION ... 70
6.2 WHOLE-LIFE COSTS ... 70
6.3 CAPITAL COSTS ... 71
6.4 OPERATING COSTS ... 71
6.5 RECORDS ... 73
6.6 OPERATIONAL PRACTICES ... 74
6.7 CONCLUSIONS ... 75
6.8 CHECKLIST ... 75

7 SUMMARY AND CHECKLISTS ... 76

7.1 INTRODUCTION ... 76
7.2 RECOMMENDATIONS FOR GOOD PRACTICE ... 76
7.3 CONSOLIDATED CHECKLIST ... 78

8 RECOMMENDATIONS FOR FUTURE WORK ... 86

APPENDIX A1 CONSULTATIONS ... 87

APPENDIX A2 COST DATA DERIVED FROM CONSULTATIONS ... 88

A2.1 CAPITAL COSTS ... 88
A2.2 OPERATING COSTS ... 88
A2.3 MAINTENANCE COSTS ... 89

APPENDIX A3 LITERATURE REVIEW ... 92

A3.1 CIRIA PUBLICATIONS ... 92
A3.2 FOUNDATION FOR WATER RESEARCH ... 93
A3.3 WATER RESEARCH CENTRE ... 93

    A3.4 BRITISH STANDARDS 94
    A3.5 WATER POLLUTION CONTROL FEDERATION (USA) 94
    A3.6 NATIONAL JOINT HEALTH AND SAFETY COMMITTEE FOR THE
          WATER SERVICE 95
    A3.7 MISCELLANEOUS BOOKS 95
    A3.8 MISCELLANEOUS PAPERS [UK] 95
    A3.9 MISCELLANEOUS PAPERS [USA] 96

REFERENCES 98

FIGURES

Figure 1.1	Proposed feedback system	14
Figure 2.1	Above-sewage pumping station	24
Figure 2.2	Submersible pumping station with a single wet well	25
Figure 2.3	Wet-well/dry-well pumping station with dry-well mounted submersible pumps	26
Figure 2.4	Wet-well/dry-well pumping station with shaft-driven pumps	27
Figure 2.5	Wet-well/dry-well pumping station with horizontal close-coupled pumps	28

TABLES

Table 3.1	Operational maintenance activities	38
Table 3.2	Planned maintenance activities	39
Table 3.3	Relative costs of MDPE pumping mains	44
Table A2.1	Design and construction costs for six submersible pumping stations, based on completed questionnaires	90
Table A2.2	Operating and maintenance costs	91

Glossary

centrifugal pump	The term is strictly applicable only to radial and (possibly) to mixed-flow pumps, but tends to be used for all types of rotodynamic pumps, including axial flow pumps.
characteristics	The pump characteristics are a set of graphs relating the various pump variables that serve to describe the pump performance. These are normally shown as pump total head rise, power, efficiency, and net positive suction head (NPSH) plotted against pump discharge as the main independent variable, separate curves being plotted for different pump speeds.
closed valve head	The pump total head rise at zero flow rate. Also referred to as shut-off head.
combined system	A sewerage system in which foul sewage and surface water drainage are carried in the same drains and pipes.
discharge	The volumetric flow rate, Q, at the pump outlet flange; measured in cubic metres per second for consistency in metric calculations. In practice, many other units are used, e.g. litres per second and megalitres per day.
dry weather flow	When the wastewater is mainly domestic in character, the dry weather flow (DWF) is the average daily flow to the treatment works during seven consecutive days without rain (excluding a period that includes public or local holidays) following a period of seven days during which the rainfall did not exceed 0.25 mm on any single day.
I P rating	A coding system included in BS 5345 and used to define the degree of protection offered by the enclosure(s) of equipment to the ingress of foreign bodies, dust and liquids.
motor control centre	A group of motor starters, each within separate compartments accommodated within a cubicle-pattern switchboard.
non-return valve	A valve installed within a pumping main to prevent reverse flow and designed to close very quickly when forward flow ceases.

positive displacement pump	A pump in which fluid is drawn into and expelled from a chamber by a piston or other means.
progressive cavity pump	A pump that has a helical rotor that revolves inside a suitably shaped stator, thereby forcing liquid from inlet to outlet.
rotodynamic pump	A machine that acts by imparting kinetic energy to the fluid by means of a rotating impeller. Pressure is generated by the subsequent conversion of the kinetic energy to static head in the casing element of the pump.
separate system	A sewerage system in which foul sewage and surface water are conveyed in separate pipes.
septicity	The anaerobic decomposition of organic matter in sewage. It will arise in stagnant sewage such as may occur in long, flat sewers and sumps, particularly in warm weather. Hydrogen sulphide gas, H_2S, is a product of the decomposition, and other foul odours can occur. In chambers and sewers, the H_2S may combine with water vapour above the liquid level to produce sulphuric acid, H_2SO_4. This may condense on the walls to attack concrete surfaces and corrode metal fittings.
sewage	The waterborne wastes of a community. It may include domestic, commercial and industrial sewage, and infiltration and stormwater, the combination depending on the nature and condition of the catchment area and sewerage systems.
submersible pump	A pump with a direct-coupled motor that can be operated submerged.
surface water	The runoff from paved and unpaved roads, buildings and land resulting from precipitation (rain or snow).
vertical-spindle pump	A pump that is connected to its motor by a long vertical drive shaft.

Abbreviations

CDM	Construction (Design and Management) Regulations 1994
DI	ductile iron
DWF	dry weather flow
GRP	glass-reinforced plastic
HAZANS	hazard analysis
HAZOPS	hazard and operability study
H&S	health and safety
ICA	instrumentation, control and automation
IP	degree of protection coding included in BS 5345
M&E	mechanical and electrical
MCC	motor control centre
MDPE	medium-density polyethylene
NPSH	net positive suction head
NRA	National Rivers Authority
NRV	non-return valve
O&M	operation and maintenance
QA	quality assurance
REC	regional electricity company
uPVC	unplasticised polyvinyl chloride

1 Introduction

1.1 DEFINITIONS

1.1.1 Buildability

The term "buildability" when applied to a design relates to its ease of construction or the "ability" of the contractor to "build" it. It also facilitates the use of safe construction methods.

The Construction (Design and Management) Regulations 1994[1] is a statutory instrument that requires the designer to consider the avoidance or reduction of risk to people's health and safety arising from construction, maintenance etc while carrying out his design. He must also inform the principal contractor of potential hazards that may be encountered during construction. In so doing, it seeks to improve buildability.

Buildability should not take precedence over other major design principles such as function, stability and aesthetics. However, it should be taken into account when considering these principles.

An important aspect of buildability is the way in which the overall design concept is developed, detailed and communicated to the contractor, and the extent to which it takes into account the practical constraints of the construction process. A formal definition of buildability was proposed in CIRIA Special Publication 26, *Buildability – an assessment*[2], as follows:

"The extent to which the design of a building facilitates ease of construction, subject to the overall requirements for the completed building."

In that publication, it was concluded that:

(a) Good buildability leads to major cost benefits for clients, designers and builders.

(b) The achievement of good buildability depends upon both designers and builders being able to see the whole construction process through each others' eyes.

The CIRIA publication *Practical buildability*[3], by Stewart Adams, develops the concept of buildability and applies it to a wide range of design examples.

1.1.2 Maintenance

The term "maintenance" covers all the operations necessary to ensure that the building and the plant contained therein are maintained in good condition and at optimum operating efficiency throughout the design life of the building. This covers everything from routine servicing to major overhauls. Since the building may have a design life of two to three times the life of the mechanical plant it contains, the term "maintenance" is extended in this report to cover major rehabilitation work incorporating the replacement and possible upsizing of plant within the existing building.

Maintenance can be planned or reactive. In the context of this report attention is focused on planned maintenance. However, similar issues, particularly the health and safety aspects, need to be considered during reactive maintenance operations.

Maintenance is not necessarily compatible with buildability. Thus in applying the principle of good buildability and optimising the construction process, it is essential to consider the maintenance operations that will need to be carried out during the life of the building and plant. Appropriate provisions need to be made for working space, means of access, lifting facilities etc., to enable the work to be carried out as safely, quickly, efficiently and economically as possible.

Thus, the best design is one that balances good buildability with ease of maintenance to provide the best overall advantage in terms of whole-life costs.

1.2 BACKGROUND TO THE STUDY

Sewage pumping stations are complex units requiring the integration of the civil (and sometimes architectural) design and construction process with the design, installation, operation and maintenance of major items of mechanical and electrical plant and equipment. It is essential that an overview be taken at the design stage to consider and balance the effects of this complexity and the aggressive operating conditions. This is particularly relevant to ease and safety of construction, operation, maintenance and rehabilitation, taking due account of the risk of accidents. Where a pumping station is difficult to operate and maintain, there is an increased risk of breakdown leading to consequential costs for damage and inconvenience.

Initial points of reference include CIRIA Report 121, *Design of low-lift pumping stations with particular application to wastewater*[4], and CIRIA Building Design Report, *Practical buildability*[3]. The former describes the theoretical background to pumping station design and gives a checklist of items to be considered to ensure buildability and maintainability of pumping stations, but no detailed guidance. The latter sets out 16 principles for good buildability, with examples, but is mainly concerned with building construction rather than civil engineering works. Improved design can result in cost savings and eliminate safety hazards at the following stages:

- building – initial construction
- operation – ease and convenience of normal operation including routine maintenance, cleaning and inspection of mechanical, electrical and control equipment
- maintenance – dismantling and reassembly in order to repair, replace or rectify a malfunction
- rehabilitation – work required only once or twice during the life of a pumping station, such as replacement of pumps and other plant and equipment and possible upsizing to increase its capacity.

So far as is practicable, the design should allow for each of these operations to be carried out quickly, easily and safely at minimum cost. Where routine operational maintenance is carried out regularly and efficiently, there should be a reduction in the frequency and scale of major maintenance and rehabilitation.

1.3 FEEDBACK

An effective experience feedback system would help pumping station designers to avoid repeating past mistakes, and to encourage the adoption of new ideas. Most companies provide the framework for feedback within their quality system, but support for making the feedback system truly effective varies throughout the industry. It depends on the commercial climate, perceived priorities at director level, and on the energy and enthusiasm of individual engineering managers and designers within the organisation. A proposed feedback system is illustrated in Figure 1.1.

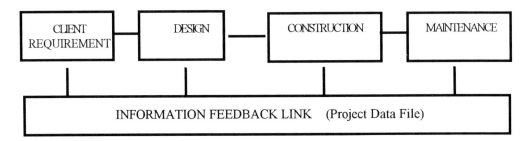

Figure 1.1 *Proposed feedback system*

Ideally, the experience of all parties (owners, operators, maintaining authorities, designers, constructors and suppliers) should be channelled into a central system. This could be based on a carefully designed database that could be interrogated by using a keyword system. Access could be via a modem and the Internet.

The ultimate beneficiaries of an efficient feedback system will be the pumping station owners, operators and maintaining authorities. It is therefore suggested that they should facilitate the collection of data on buildability and maintainability and make that information available to the designers.

1.4 SCOPE OF THE STUDY

The research covers the most common types of sewage pumping station constructed during the last 25 years. The aim of the study is to produce a manual of practical guidance for use in design offices, the application of which will reduce the whole-life costs of sewage pumping stations by taking into account the ease, speed and cost of their construction, operation, maintenance and rehabilitation. The manual is intended for general use, providing guidance in relation to conceptual and detailed design. If a particular aspect is only applicable to, or exclusive to, certain types of pumping station, this is stated at the relevant point in the report text.

Advice and guidance is given on design to minimise overall capital, operational and replacement costs in relation to:

- civil engineering, building and related works
- mechanical equipment and services
- electrical equipment and services
- instrumentation to monitor, control and automate
- layout requirements for ease and speed of operation and maintenance, and related safety issues
- corrosion problems, with particular reference to reliability of mechanical, electrical and control equipment in damp and aggressive conditions
- maintenance strategies.

The following topics are covered:

- communication between designer, contractors and operators via drawings, specification, operational manuals, meetings etc.
- obtaining economic advantage by speeding up construction
- proper access to the pumping station site
- standardisation within and between projects to reduce costs
- prefabrication to minimise on-site work
- avoidance of complicated and difficult detail which, while being structurally efficient in terms of reducing material costs, may add significantly to overall construction, operation and maintenance costs
- reuse of formwork during construction
- standardisation of working practices to speed construction and maintenance
- internal access and space requirements for maintenance and auxiliary equipment
- pump set selection
- equipment selection and maintenance requirements (valves, lifting and cables etc.)
- instrumentation and control equipment selection and installation
- standardisation of equipment
- pump condition monitoring
- malfunction
- standby facilities and other emergency provisions
- standard packaged pumping plants.

The research does not cover:

- the rehabilitation of old pumping stations
- pumping stations for purposes other than sewage pumping
- design of equipment and components for ease of maintenance and operation
- pumped storage tanks (i.e. where the wet well is of considerable size in relation to pump capacity. These are considered to be a special case.

1.5 STRUCTURE OF THE BOOK

The book has been structured to provide guidance with background information for improved buildability and maintenance of sewage pumping stations.

Section 2 Construction and plant installation

Covers the civil engineering aspects of pumping station design and construction. It deals with site constraints, station size and layout, and scope for precast or package units; mechanical, electrical plant and control systems or equipment; provision of services, safety matters, types of contract and the relationship between the designer and contractor with particular reference to the division of responsibilities and effective communication.

Section 3 Routine operation and maintenance

Covers operation and routine maintenance. It deals with sump design, economic pumping main design, hydraulic design including surge analysis, control systems, access for maintenance operations, safety and reliability.

Section 4 Major repairs and maintenance

Covers major maintenance and repair. It deals with methods of maintaining flows, facilities for access, plant removal and dismantling, design of pipework and power connections for easy and safe removal and replacement of plant and equipment. It also considers methods of coping with vandalism.

Section 5 Rehabilitation

Considers the problems of rehabilitation, particularly where increased capacity is a requirement, and aims to provide assistance in foreseeing these problems and making reasonable allowances in the initial pumping station design.

Section 6 Balancing whole-life costs

Considers the interrelation of various cost elements of pumping station construction, operation, maintenance etc., and how to balance them to give optimum whole-life costs. Operation and maintenance costs are often difficult to estimate, and information from pump suppliers needs careful interpretation (the keeping of adequate records can provide valuable help in this respect).

Section 7 Summary and checklists

Summarises the findings in the earlier sections, and makes recommendations for good practice and improved QA procedures. Design checklists are provided for use in association with quality assurance procedures as a check that recommendations have been properly considered in the design process.

It is envisaged that most practitioners will initially read Sections 1 to 7 and subsequently go straight to Section 7 when carrying out or checking pumping station design work.

Section 8 Recommendations for further work

Recommends standardisation of record-keeping and specifications to assist in improving certain aspects of existing practice identified during the course of the study.

2 Construction and plant installation

2.1 INTRODUCTION

This section considers design for ease, speed and economy of construction. By their very nature, these aspects are primarily concerned with civil and building works, which usually account for the largest initial capital outlay for most pumping stations. However, the design of the civil and building works should also take into account the space and access requirements for construction and installation of mechanical plant, and electrical and control equipment. The final design should be the product of all the relevant disciplines.

This report covers the various duties of the "design team" and is intended to cover all probable contract arrangements, whether they follow the traditional pattern or the increasingly popular design and build, or design, build and operate types of contract. All the functions of the "team" will be required irrespective of whom the various team members are working for.

2.2 GENERAL DESIGN CONSIDERATIONS

The requirements for a small pumping station serving a small group of isolated properties are different from those for a pumping station serving a reasonably large housing estate. In turn, the latter's requirements will be different from those for a station serving a sizeable community, or a terminal pumping station transferring the flow from a whole town to a sewage treatment works some distance away.

It is not the role of the publication to provide comprehensive guidance on the "functional" design of pumping stations, and although the following aspects impact on design for buildability and maintainability they are not treated in detail. Such guidance should be obtained from *Design of low-lift pumping stations*[4].

The design of a sewage pumping station forms part of the design of the overall sewerage system. It should be recognised that the civils structure has three primary functions:

1. to contain and manage the sewage flows
2. to accommodate, support and protect the mechanical plant, together with the electrical, instrumentation and control plant and equipment
3. provide access and support to facilitate the installation, maintenance and replacement of plant and equipment.

The following factors have an effect on the design of the pumping station, the plant to be installed in it, and the complexity of the control system required.

General

- availability and cost of power supplies
- risk of flooding
- allowance for extensions to accommodate possible future development and increases in flow.

Function

- whether the upstream sewage flow system is combined or separate. There may be a potential problem with grit
- potential constraints on the provision of an overflow (if applicable)
- need for screening for pump protection and/or of overflows
- length of pumping main
- need for ventilation and/or odour control
- external communications
- protection from vandalism
- security and fire alarms
- statutory requirements and planning permission.

Construction

- location and depth of upstream sewer
- location of pumping main discharge, its elevation relative to the pumping station and any constraints on the rate of discharge
- availability of land
- ground conditions (thorough site investigation of what is likely to be relatively low-lying land)
- aesthetics (i.e. its size, location and local planning requirements).

Plant and instrumentation

- access for personnel and plant.

The following issues need to be considered in the design of the pumping station and suitable precautionary measures taken as appropriate:

- whether the sewerage system is combined or separate. The variation of flows is much greater in a combined system, and storm overflows, storm tanks and/or increased sump storage may be required
- risk of a pollution incident in the event of a mechanical or electrical failure
- risk of flooding, if no overflow provided. Standby power generation or emergency call-out procedures triggered by an alarm may need to be provided
- in a combined sewerage system, significant quantities of grit may arrive at the station during and after storms.

Economic considerations have an effect on the pumping station design:

- ease, and hence cost, of maintenance will affect the design
- in terms of whole-life costs, a wet-well/dry-well pumping station may be more economical than a submersible station, particularly if large pumps are used

- the relationship between velocity and head should be considered: a smaller-diameter pumping main will cost less to provide and lay, but the higher velocity in the main and consequent higher head on the pumps will increase operating costs, and may require a larger or different type of pump and/or surge protection. Thus the most economic size of pumping main will not necessarily be the smallest practicable
- there are exceptions to the above point, and the following example demonstrates some of the various issues that need to be considered.

For a small, isolated group of houses, the length of pumping main may make the connection of these properties to the main sewer uneconomic unless a small-bore flexible pipe can be laid cheaply by mole plough. This would determine the type of plant required to cope with the relatively low flows together with the significantly higher heads generated by the higher velocities in the smaller-diameter main. Both positive displacement and submersible pumps are suitable for these duties. A positive displacement pump with maceration needs to be located above the top sewage level and should be provided with a pressure relief valve for safety reasons. Normally, for convenience of construction and access for maintenance, these pumps are located at ground level and provided with a small building. Submersible pumps, employing proprietary grinders or cutters, would enable the station to be located below ground with no superstructure.

Statutory requirements with implications for the design and construction of sewage pumping stations would include (but not be limited to) the following:

- Construction (Design and Management) Regulations 1994[1]
- Control of Substances Hazardous to Health Regulations (COSHH) 1994[7]
- Electricity at Work Regulations 1989[8]
- Environmental Protection Act 1990[9]
- Health and Safety at Work Act 1974[10].

Depending on its location, the planning authority may apply quite stringent conditions to the planning consent, which may have a significant effect on the design of the pumping station.

The following general design considerations are applicable to all pumping stations:

- **adequate** room for **access** to pumps, motors, valves etc. for inspection, maintenance and replacement of plant
- **lifting equipment**, davits, lifting beams with block and tackle or powered lifting equipment should be provided for ease of installation or removal of heavy plant and equipment. Alternatively, a removable cover may be used to provide for access by a mobile crane. It may be appropriate to have a combination of permanent lifting gear, for maintenance, and a mobile crane for complete items during replacement or rehabilitation
- **adequate lighting** should be provided for inspection and maintenance of all equipment; where such equipment is installed in wet wells, it should be of an appropriate Ingress Protection (IP) rating defining the resistance to water and gas penetration and degree of explosion protection. For smaller pumping stations, where no provision is made for personnel access into the wet well and

pumps are lifted out for inspection and maintenance, externally mounted floodlighting might be more suitable
- a **water supply** for washing and hosing down the installation
- the **overall internal layout** should be logical and tidy. This not only facilitates the maintenance of equipment and of the internal structure, but also encourages operations staff to keep the station in a workmanlike condition, and provides a safe working environment.

2.3 HEALTH, SAFETY AND WELFARE DESIGN

The Construction (Design and Management) Regulations 1994 (CDM)[1] require the designer to take account of all foreseeable risks to the health and safety of any person at work, relating not only to construction but also to the subsequent operation and maintenance of the installation.

When designing a pumping station it is essential to comply with these statutory requirements. In addition, there may be other relevant considerations, or specific requirements from clients.

Typically, the following need to be considered:

- **confined space** working and entry restrictions
- **hazardous area** assessment and designation
- **skin/eye contact** with sewage – avoidance and aerosol effects
- **safety handrailing** and similar barriers
- **ventilation** and odour abatement
- **lifting equipment** – portable or fixed
- **adequacy of space** for operation and maintenance
- **access** into chambers
- **external roadway** access to the station
- **messing** facilities
- **control and monitoring equipment**
- **general and emergency lighting** appropriate for the type of pumping station
- **slips, trips and falls**
- **off-site communication**.

2.4 CONSTRUCTION METHODS

With the aim of improving buildability, the designer should consider holding consultations with civil contractors regarding construction methods, working space and access, particularly where space is restricted or difficult ground conditions are likely to be encountered.

Depending on the location of the pumping station and the ground conditions, the civil designer may need to give careful consideration to suitable construction methods and to design the station accordingly. So far as possible, the contractor usually determines the construction method, subject to the approval of the engineer, although, occasionally, it may be necessary to specify the method the contractor should adopt.

Examples of concerns that may have to be addressed are given below:

- as the station may be located at a low point, there is an increased likelihood of high groundwater levels requiring dewatering during construction and with greater potential for occasional flooding
- available land may be unsuitable for other uses because subsoil conditions are poor, or it contains fill material
- site may be cramped, with restricted access and working space in a potentially hazardous location
- limited storage space for materials etc.
- requirements relating to the architectural style of the superstructure
- special requirements in respect of noise and other nuisance.

These considerations need to be taken into account by the designer if he is to produce a suitably "buildable" design. It would be prudent to discuss these with a potential contractor at an early stage as possible, and to be reasonably flexible if the appointed contractor wishes to follow a different methodology.

Buildability can be further improved by reducing potential construction hazards. The contractor has obligations under CDM to complete the safety plan and to carry out risk assessments; an early input by a contractor may enable the designer to eliminate as many of these risks as possible.

Some of the above considerations, such as restricted access, will affect the ease of maintenance of the plant, and special provisions may need to be considered.

Important details to be considered at the design stage are described below:

- for underground structures, flotation is a major design consideration. If the groundwater level in the surrounding soil is high and the structure itself is dewatered, there will be a large hydrostatic upthrust on the structure. To prevent flotation under these conditions, it must be resisted either by increasing the weight of the structure or by external restraint
- the detailed design of the connections between the incoming and outgoing sewers and pumping mains and the main structure requires special consideration in relation to the foundation conditions. Inadequate provision for expansion/contraction and for differential settlement can lead to leaks at locations where access is difficult after construction is complete and can lead to foundation damage
- the construction drawings should, where necessary, allow some flexibility, such as extra space, box-outs etc., where mechanical and electrical details are unresolved at the time of construction
- the construction of the underground concrete work represents the major part of the civil works costs. The choice of sump size, particularly its depth, and whether to have a single wet well or a wet-well/dry-well arrangement, are therefore crucial decisions that should be made after optimising the hydraulic and operational requirements
- the most appropriate method of construction must be selected; options range from complete package sump and pumping equipment simply positioned on a prepared foundation, through construction using precast concrete sections, to cast-in-situ concrete. For all but the smallest pumping stations, in-situ construction is still the most widely used method. However, the type and size

of standard precast concrete units is increasing and careful consideration should be given to their use to speed construction. The precast units can range from single shaft or caisson ring units to segmental shaft sections.

Many of the design issues described above also affect buildability, and consequently merit particular consideration. By their very nature, many of these issues are interrelated. The principal points that most frequently need to be addressed are:

- *Site conditions* A high groundwater level and/or poor soil conditions can give rise to excavation problems. Groundwater must be controlled and foundation conditions may need improvement; the methods adopted will depend on soil conditions, groundwater levels and the depth of excavation. Techniques to be considered include: dewatering; sheet piling; caisson construction; diaphragm walling; contiguous bored piling; ground freezing/treatment. The appropriate technique dictates, or is dictated by, the pumping station type and layout.
- *Excavation* Sinking a pumping station as a conventional or jacked caisson (using prefabricated rings or segments, or in-situ reinforced concrete construction) may be more economic than an open excavation, but this will depend on size, soil conditions, access and final configuration. The standard shape is usually circular. Sheet-piled temporary structures or diaphragm walling are often required for larger stations, in order to maintain an open excavation while permanent construction is undertaken inside; parts of these may be incorporated into the permanent works.
- *Shape* A cylindrical design for the outer walls of the substructure instead of a rectangular design is attractive both structurally (because a cylinder is an efficient structural form) and also from a construction viewpoint (because the temporary works can also act as shuttering, particularly when prefabricated rings or segments are being used). This can produce a lower cost for a major part of the whole pumping station, although there may be extra costs from more complex internal arrangements that are often needed to provide good hydraulic conditions.
- *Standardisation/rationalisation* Standardisation of components and sizes should be considered to gain benefits from multiple reuse of temporary formwork, for example, and using precast components to ease and speed the construction process, particularly with regard to the substructure. Structural forms, especially first-stage concrete shapes below ground should be rationalised, with any complex internal layouts and benchings being constructed in subsequent second-stage concrete pours.
- *Detailing/specification* Specified construction tolerances should be considered against final structural and hydraulic requirements, and the anticipated method of construction. The detailing of reinforcement should promote the ease and speed of fixing as a means of reducing the construction time for substructure works. The need for complex construction joints should be minimised and, if required, attention paid to waterproofing. The specification and monitoring of the quality of the concrete work needs to have regard to the strength, watertightness and durability of the finished structure: consideration should be given to limiting the number of concrete grades specified for any individual structure.
- *Aesthetics* Sewage pumping stations can often be the subject of public prejudice and attract attention by their very nature. The positions of any above-ground works and structures and their architectural design, including

materials and colour, should therefore be in sympathy with the local environment. Attention to these matters should in turn also address buildability issues, many of which are highlighted in the CIRIA Building Design Report, *Practical Buildability*[3].

- *Temporary works* During construction and installation works involving rehabilitation of existing pumping stations, existing wastewater flows will need to be maintained. Wherever possible the most efficient method will be to divert flows or overpump, so that construction work can take place in isolation from sewage flows. Where this is not possible, temporary works and restricted working practices will need to be considered, with method statements and risk assessments in accordance with the CDM Regulations.

2.5 PLANT SELECTION

The type of pumping station will depend mainly on the size and number of pumps required, the size of pumping main, the maximum pumping head, and operation and maintenance requirements.

Pumping stations can be considered as falling into the following categories:

- ejector stations ("package" type or constructed in situ)
- submersible pumping stations ("package" type or constructed in situ)
- positive displacement pumping stations (with/without macerators)
- wet-well/dry-well pumping stations. (There is a variety of possible pump arrangements, including dry-well-mounted submersible pumps, vertical-spindle centrifugal pumps, close-coupled or with motors at ground level and horizontal spindle pumps. Fully equipped "package" dry wells are available, which can draw from a separate sump constructed from manhole rings.)
- Archimedean-screw pumping stations.

Recommendations on actual pump selection are beyond the scope of this report. Details can be found in standard texts and CIRIA Report 121, *Design of low-lift pumping stations*[4].

Ejector stations are small pumping units suitable for housing estates; however, they are rarely used these days and are not included in *Sewers for adoption*[11]. Although there are still many installations of this type operating satisfactorily, they are not discussed any further in this book. Similarly, there is no further discussion of screw pumps, as they have a special application and are rarely used within foul or combined sewerage systems.

The advantages and disadvantages of the most common types of arrangement for sewage applications are summarised in the following sections. In all cases, consideration needs to be given to selection of electrical equipment appropriate to the hazardous area classification.

2.5.1 Above-sewage pumping station

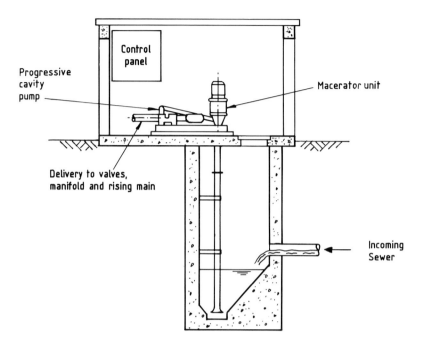

Figure 2.1 *Above-sewage pumping station*

The advantage of above-sewage pumping stations is that the pumps are housed in a dry enclosure, out of the sewage, with access at ground level. Normally, progressive cavity pumps are used in this application. The plant for smaller stations is available as a factory-built system that can be contained in GRP or similar enclosures.

Installation can be a matter of positioning the pump over a sump and connecting the relevant services. The pump needs to be kept primed and, unless the pumps are self-priming, a foot valve will be required. There are consequential dangers associated with dry running. The low pump speeds in the latest generation of progressive cavity pumps help to reduce wear (particularly relevant where grit is present) and allow for long periods between routine maintenance, which, when required, can be carried out on site in a clean environment. Progressive cavity pumps are more hydraulically stable (by maintaining a constant discharge over a range of delivery heads at fixed speed) than centrifugal pumps, can pump against higher heads, and are generally more efficient. The disadvantage of a progressive cavity pump is its relatively low volume capability. Therefore, applications at high flow and/or low head may require an economic appraisal.

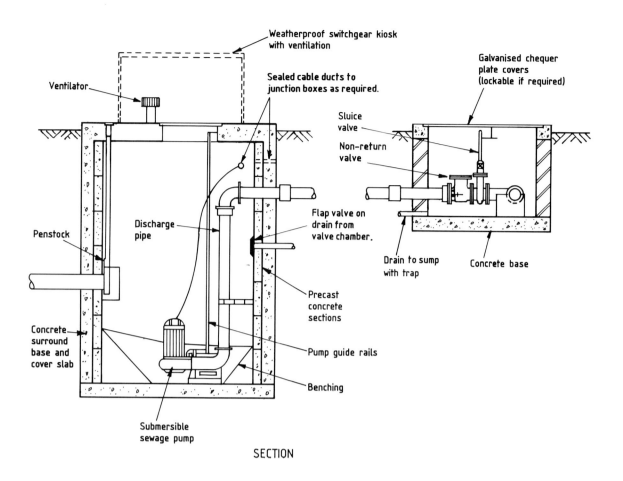
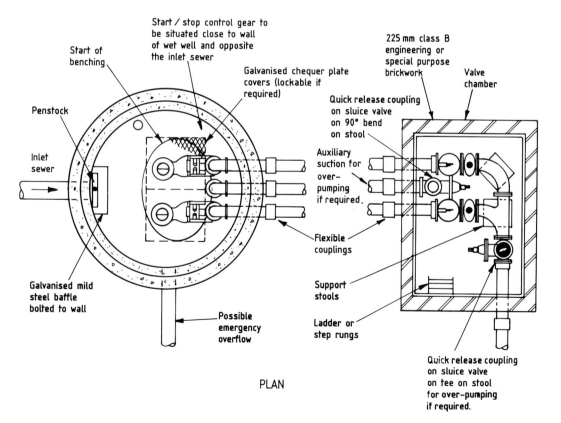

Figure 2.2 *Submersible pumping station with a single wet well*

2.5.2 Submersible pumping station with a single wet well

The advantages of these one-cell pumping stations are that they are relatively simple and cheap to construct, and can be made unobtrusive (usually only the control kiosk is visible above ground). The smaller pumping stations can be purchased in self-contained rotationally moulded polyethylene or GRP packaged units, while the bigger stations are usually constructed on site using precast sections with concrete surround.

The main disadvantage of wet-well submersible pumping stations is that the pumping equipment is fully or partially submerged in the sewage and has to be lifted out for inspection and maintenance. While this operation is reasonably simple with appropriate lifting equipment, and should not be required very often, the pump requires washing down before maintenance can be carried out.

For ease of operation and maintenance, a separate valve chamber is often provided external to the wet well. Emergency overpumping connections can be provided in this chamber.

Macerating pumps are available for situations where there is likely to be a problem with rags etc.

2.5.3 Wet-well/dry-well pumping station with dry-well-mounted submersible pumps

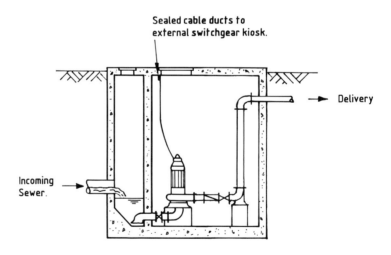

Figure 2.3 *Wet-well/dry-well pumping station with dry-well-mounted submersible pumps*

This type of pumping station has the same advantages as the wet-well submersible type, but the pumps and pipework are in a dry environment; therefore, inspection and routine maintenance are simpler and safer, since it is not normally necessary to lift the pump out to do this. Also the non-return valve can be located adjacent to the pump. The dry-well configuration is normally used when inspections and routine preventative maintenance are likely to be carried out more frequently.

The disadvantages of this type are that the two-cell structure is more expensive and that the dry well may need to be treated as a confined space, depending on its hazardous area classification. Adequate ventilation will be required for access purposes. In addition a small sump drainage pump is required to remove any leakage from pump and valve glands. The dry well needs to be big enough for man-entry and to facilitate maintenance operations. Consideration also needs to be given to the cooling of submersible pumps in this type of arrangement.

This configuration is usually used to avoid the need to locate the motor above ground level (thus no superstructure is required) and to cover the possibility that the dry well could flood. The dry well is usually drained using portable pump sets when necessary.

2.5.4 Wet-well/dry-well pumping station with shaft-driven pumps

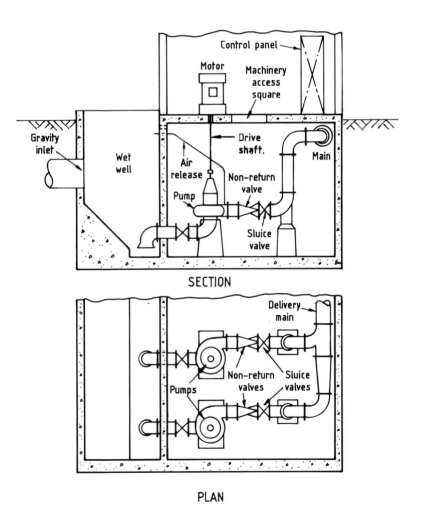

Figure 2.4 *Wet-well/dry-well pumping station with shaft-driven pumps*

This type of pumping station is of the traditional design with a superstructure housing pump motors and switchgear mounted above flood level. Depending on its size, lifting equipment, store rooms, workshop, toilet and washing facilities may be provided. It offers a dry, covered area for repair and maintenance (particularly for the electric motor and switchgear) and other operational facilities for staff at large pumping stations.

The disadvantages are the same as for the wet-well/dry-well type with dry-well submersible pumps. The shaft drives require structural support for any intermediate bearings, access for maintenance and safety guards. Where fixed lifting equipment is not installed, portable davits or other mobile lifting equipment will be required for routine inspection and maintenance. Additional security is needed for the above-ground building. This form of pumping station is now less popular due to its higher capital cost than the alternatives, although it is still commonly used for larger pumping stations.

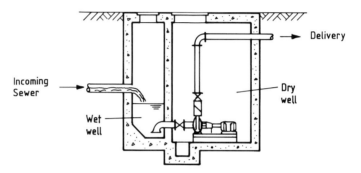

Figure 2.5 *Wet-well/dry-well pumping station with horizontal close-coupled pumps*

There are many existing wet-well/dry-well pumping stations in operation with either vertical- or longitudinal-spindle close-coupled pumps (Figure 2.5). These would not normally be provided in new installations.

2.5.5 Small pumping installations

As mentioned previously, a group of houses a long distance from the main sewer may be served by an above-sewage pumping station incorporating a macerator and a positive displacement pump (e.g. progressive cavity) delivering into a small-bore pumping main. The sump need be no more than a specially designed manhole with the pump and ancillary equipment located in a small enclosure at ground level. This type of pumping station can often be disguised by installing it in a garage; the vehicle door provides good access for maintenance. Alternatively, a packaged submersible pumping station may be more suitable and, if required, submersible grinder/macerator pumps can be used. However, the choice of pump will depend on the flow/head characteristics of the system.

Several operators consider that single wet-well submersible pumping stations should be provided only when the capacity of each pump is less than 350–400 l/s. Where larger pumps are employed, a wet well/dry well pumping station would be preferred, to enable inspections and routine maintenance operations to be carried out without lifting the heavy pumps out of the sump. It may be preferable to avoid providing fixed lifting equipment in remote locations because of the cost of regular inspection and statutory testing. The use of mobile cranes should only be necessary for occasional major maintenance and repair. However, other operators favour submersible sewage pumping stations, employing submersible pumps having capacities well in excess of 400 l/s.

Care needs to be taken with deep submersible stations with relatively shallow valve chambers. The non-return valve (NRV) should not be more than about 8 m above bottom water level in the wet well, as, when the pump is not operating, very low pressure or a vacuum will occur adjacent to the NRV due to the column of water between the NRV and the pump. This will tend to cause gases to come out of solution and create an air-lock. In such circumstances, consideration should be given to the use of an air valve to permit column draindown.

2.6 SUPERSTRUCTURE DESIGN

The purpose of the superstructure is to provide weather protection and security for the equipment and environmental comfort for operators. The design of the actual building requires special consideration in respect of its application, purpose, size, type and aesthetic appearance.

Other points to be considered are:

- general access
- lifting requirements, adequate openings and working areas for plant installation, maintenance and replacement
- lighting, heating and domestic power arrangements
- ventilation
- odour control
- protection against vandalism
- security and fire alarms
- flooding risk
- lightning protection
- noise attenuation.

2.7 WET WELL DESIGN

Virtually all wastewater pumping duties require a suction sump to receive the incoming flow and to provide a storage volume to accommodate pump output to inflow rate variations by means of "on-off" operation, either alone or in conjunction with speed control. The "on-off" operation of the pumps is normally effected by means of level-monitoring devices in the sump. Non-intrusive devices such as ultrasonic level monitors are usually preferred, as they do not collect rags or fats that would inhibit their operation. The following points should be considered in sump and pump chamber design:

- In larger pumping stations, the sump may be divided into two or more compartments to allow individual cleaning and execution of possible future rehabilitation work. In order to accommodate this flexibility, it is necessary for the incoming sewer to be divided and valved to the separate compartments.

- The partition wall should be provided with a penstock, which is kept open during normal operation.
- The sump floor should slope so as to promote uniform distribution of flow and conveyance of solids to pump suctions, preferably by means of steep benching to the floor. Baffle walls may also be needed to dissipate the energy in the incoming flow.
- The depth of the sump should be such that the highest normal operating or "start" level is below the invert of the incoming sewer, to avoid backing up, unless provision has been made for this in this design of the sewerage system.
- For large wet wells, particularly if not of conventional design, careful consideration should be given to model testing. The aim of this is to ensure that the hydraulic characteristics are satisfactory, i.e. to avoid flow conditions at the pump suctions such as vorticity and pre-swirl, or formation of excessive deposits of grit and rags. Poor sump design can have a major adverse effect on pump performance and reliability.
- An emergency overflow may be provided as the most appropriate means of preventing flooding during a power failure (alternatives may be dual power supplies or standby generation facilities). This depends on there being a suitable discharge point nearby and the agreement of the NRA, whose requirements should be complied with. There may also be a storm overflow associated with the pumping station that will operate under particular rainfall conditions. It is normal for the NRA to require that precautions be taken to ensure that rags and other unsightly materials are retained in the sewage flow and not discharged into the watercourse. There is a variety of screening plant available for this application. The NRA requirements for overflows can be very stringent and the screening facilities required can have a significant impact on the sump size and arrangement. This needs to be known at an early stage in the design.

Increasingly, there is a requirement to provide storage capacity in the sump, in order to balance flows. This is usually linked with a restricted capacity in the downstream sewers and/or a restriction in the frequency of operation of any storm overflow that may be sited in the vicinity of the pumping station. Long retention periods in the sump can have a significant impact on the operation of the pumping system through deposition of solids, increased septicity etc., and considerable care needs to be exercised in its design.

Wet-well design is one of the most critical elements in pumping station design, and many operational problems are attributable to unsatisfactory inlet conditions. This subject is covered in some detail by M J Prosser in *The hydraulic design of pump sumps and intakes*[12].

2.8 SERVICES

Consideration needs to be given to the following:

- availability and cost of power supply
- starting currents, electrical protection and discrimination
- standby power requirements (dual supply, standby generator: fixed or portable)
- configuration and size of switchgear and type of enclosure
- location of and ease of access to switchgear and control equipment
- cable routing and accessibility
- type, location and accessibility of services (e.g. heating, lighting, fire and intruder alarms, socket outlets and local plant controls)
- the need for "zoning" and explosion rating in respect to electrical equipment. *Sewers for adoption* (4th edition)[11], which is concerned primarily with small submersible pumping stations, recommends that pump motors should be suitable for use in Hazardous Zones 1 or 2 as designated by the undertaker
- adequate ventilation to dissipate heat from electric motors and variable-speed drives
- odour control
- water supply
- lightning protection to the structure and susceptible electronic equipment
- telephone, radio or land-line communications.

2.9 COMMUNICATION

The design and construction of a pumping station is the product of teamwork involving the designer, the end user/operator, the plant and equipment manufacturers, the construction and installation contractors etc. In the context of buildability, two levels of teamwork are particularly important:

1. teamwork between civil and mechanical/electrical engineers, and
2. teamwork between designers and constructors.

While almost everyone will pay lip service to this ideal, all too often it is ignored. The result may be a technical specification that the contractor considers to be unreasonable and which precludes the use of standard items and systems for no apparent reason, an inefficient pumping station layout, and construction difficulties due to inadequate working space or access. It may also lead to increased plant costs and subsequent difficulties and delays in obtaining replacement parts.

Consultation with manufacturers/contractors will enable the designer to allow for the inclusion of standard items (where practicable) with associated cost and maintenance advantages, and ensure that the design does not incorporate any requirements that are unnecessary or that may incur excessive costs.

The designer should allow tenderers for plant contracts to point out defects in the design, either as tender qualifications or by offering an alternative design. There are examples of pumping stations, such as deep submersible pumping stations, where special provisions are required to avoid potentially serious operating and maintenance problems. Heeding the advice of the pump manufacturer could significantly reduce costs.

With the aim of improving buildability, the designer should consider holding consultations with civil contractors regarding construction methods, working space and access, particular where space is restricted or difficult ground conditions are likely to be encountered.

The nature and form of the above-mentioned consultations will vary considerably according to the type of contract to be employed and who is responsible for co-ordinating inputs. It will depend on whether the contract is of the traditional type (where the detailed design is carried before inviting tenders from civil and plant contractors), design and construct, or some other form of contract procurement.

The aim should be to produce a simple and appropriate specification setting out the required performance and design life of the plant and key components, together with the general obligations of the contractor. The contractor should then take full responsibility for the detailed design of his plant and for compliance with the specified performance requirements.

2.10 CONSTRUCTION (DESIGN AND MANAGEMENT) REGULATIONS[1]

A major aspect of buildability is to protect the health and safety of construction operatives, while safe access and adequate lifting facilities contribute to ease of operation and maintenance. Proper compliance with the CDM Regulations will make a major contribution to the elimination of many of the problems outlined in this book.

The CDM Regulations came into operation during the preparation of this report; therefore, all the pumping stations mentioned in the replies to questionnaires and during consultations were constructed before they came into force. It is too early to ascertain their impact on buildability, but it is anticipated that compliance with these Regulations will greatly improve communication between the parties involved in the construction and operation of the pumping station.

Under the CDM Regulations, there are five key parties involved in a construction project: the Client (or owner), the Planning Supervisor, the Designer, the Principal Contractor, and other contractors or self-employed workmen.

When a developer constructs a pumping station for adoption, he should be regarded as the client (or owner) until it is adopted by the water utility company, which then becomes the owner. The developer may also fulfil the role of the designer, unless he decides to employ a third party (e.g. consulting engineer) to take on that responsibility.

The key duties of the client include selecting and appointing a competent planning supervisor and providing him with information relevant to health and safety on the project; selecting and appointing a competent principal contractor as soon as it is practicable; ensuring that construction work does not start until a satisfactory health and safety plan has been prepared; and ensuring that the health and safety file is available for inspection, after the project has been completed.

If the client is appointing the designers and contractors, he must also satisfy himself that they are competent and have allocated, or will allocate, adequate resources to enable them to comply with their duties under the Regulations. (Anyone else appointing designers and contractors has similar obligations.)

The planning supervisor must co-ordinate the health and safety aspects of the project design and the initial planning. This includes, among other things, ensuring that the designers give due weight to safety considerations when preparing their designs and provide relevant health and safety information. The planning supervisor must also ensure that designers co-operate with each other, and must advise relevant parties, if required, on the competence of proposed designers and contractors and on the resources required to enable them to comply with health and safety requirements.

Under the CDM Regulations, designers must ensure that the design process takes account of the health and safety of those who are going to construct, maintain or repair a structure. Where a building contains plant, this must also cover the installation, operation, maintenance and repair of the plant. The designer's key duties include the following:

- consider, during the development of the design, the hazards and risks that may arise to those constructing and maintaining the structure, which in the case of pumping stations includes the installation, operation, maintenance and repair of the plant and equipment
- design so as to avoid risks as far as possible; to reduce them at source if avoidance is not possible; or, if it is not possible, to avoid risks or reduce them to a safe level, to consider measures that will protect workers
- co-operate with the planning supervisor and any other designers employed in the project.

One of the key duties of the planning supervisor is to ensure that a health and safety file is prepared containing all relevant information and is delivered to the client on completion of construction. This will cover many of the health and safety aspects of operating and maintaining of the pumping station and will also contain information that will assist in the planning and design of any future rehabilitation work.

2.11 CHECKLIST

The following design checklist identifies buildability and other issues in the form of points that should be addressed.

Action	Reason	Section ref.	Checked	
General				
Determine the proposed design horizon for the civil structure	Whole-life cost	2.2	/	/
Consider site constraints on the type of pumping station, e.g. access, topography, aesthetics, etc.	Layout	2.2	/	/
Consider whether a package-type pumping station would be appropriate	Small compact installation	2.5.5	/	/
Evaluate the aesthetic requirements of the locality	Statutory/environmental requirements	2.2	/	/
Decide whether or not a superstructure is required	Type of installation	2.5.3/4	/	/
Function				
Consider the type of sewerage system – separate or combined	Characteristic of flow	2.2	/	/
Determine the present dry weather flow	Design basis	2.2	/	/
Estimate the dry weather flow at the design horizon for the plant	Whole-life cost	2.2	/	/
Calculate the probable changes in flow conditions between the design horizon of the plant and the design horizon for the civil structure	Whole-life cost	2.2	/	/
Quantify the maximum flow to be pumped	Plant selection	2.5	/	/
Estimate the probable maximum flow to the pumping station	Plant selection	2.5	/	/
Determine the elevation of the discharge point compared with the pumping station	Determine static head	2.2	/	/
Measure the flow velocity in the pumping main	Determine friction head	2.2	/	/
Calculate the size and length of the pumping main and its discharge point	Plant selection	2.5	/	/
Determine the static lift required and the total head on the pumps	Plant selection	2.5	/	/
Assess which types of pumps are suitable for the required duty	Statutory requirement	2.2	/	/
Decide whether a storm or emergency overflow is necessary and where it will discharge	Services	2.8	/	/
Assess what power supply is available and its characteristics in respect of permissible starting currents	Services	2.8	/	/

Action	Reason	Section ref.	Checked	
Consider the most appropriate way of safeguarding the power supply, as well as the noise and exhaust gas emission implications of any standby power generation required	Services	2.8	/	/
Decide on the cable routing	Services	2.8	/	/
Determine the availability of water supplies	Services	2.8	/	/
Investigate the requirements for heating, lighting and ventilation, socket outlets, intruder alarms, lightning protection	Services	2.8	/	/
Determine the need for, and method of, grit removal and disposal at the pumping station. If possible, grit should be retained in the sewage flow and separated out at the sewage treatment plant	Construction Plant selection	2.4 2.5	/	/
Establish the requirements for odour-control measures at the pumping station (particularly if grit removal is carried out)	Environmental requirements	2.2	/	/
Evaluate the telemetry requirements	Communication	2.9	/	/
Buildability				
Decide on suitable locations for the site in relation to the gravity sewer and the pumping main	Layout	2.2	/	/
When choosing the location, consider access for construction plant and equipment, and the delivery and storage of plant, equipment and materials	Layout	2.2	/	/
Ensure that a suitable power supply is available	Services	2.8	/	/
Consider the ground conditions at the site, whether there is a high water table and whether the site is prone to flooding	Construction technique	2.4	/	/
Consider the sewage flow implications during construction	Construction safety	2.4	/	/
Examine the likely advantages of using standard precast or prefabricated items, e.g. precast concrete segments for sump construction, and prefabricated GRP buildings for the superstructure or control kiosk	Whole-life costs	2.2	/	/
Assess the likely benefits of adjusting the design in order to accommodate manufacturers' standard items of plant and equipment	Whole-life costs	2.2	/	/
Statutory requirements				
Ascertain whether a consent is needed for an overflow	Statutory requirements	2.2 2.7	/	/

Action	Reason	Section ref.	Checked	
Establish whether an overflow would need to be screened to minimise discharge of solids	Statutory requirements	2.2 2.7	/	/
Apply for planning permission	Statutory requirements	2.2 2.7	/	/
Examine the conditions applied by the planning authority to the development	Statutory requirements	2.2 2.4	/	/
Appoint a planning supervisor as required under the Construction (Design and Management) Regulations[1]	CDM	2.10	/	/
If the pumping station is to be adopted on completion, take into account the required design standards – e.g. *Sewers for adoption* 4th edition, or water service company design specifications	Statutory requirements	2.2 2.10	/	/
Communication				
If the pumping station is to be adopted on completion, ensure the water service company is consulted at an early stage	CDM	2.9 2.10	/	/
Initiate discussions with the following: the plant operator the pump manufacturer the M&E installation contractors the civil contractor	 Maintenance Construction Plant selection	 2.9	/	/
Ensure the CDM communication requirements have been satisfied	Statutory requirement	2.2	/	/

3　Routine operation and maintenance

3.1　INTRODUCTION

This section covers normal operation and routine (i.e. frequent) maintenance operations, such as cleaning out sumps, checking that valves open and close freely and general inspection regarding hygiene and safety. Less frequent operations that would also be covered under this heading include the regular inspection of pumps, motors and control gear to ensure that they are free from excess wear and in good operational condition. These issues require careful consideration at the design stage.

3.2　RECOMMENDED MAINTENANCE VISITS AND ACTIVITIES

The WSA/FWR *Guide to sewerage operational practices*[6] sets out the typical routine operating and maintenance activities that are given in Tables 3.1 and 3.2.

It is important that the design of the pumping station layout ensures that inspection, cleaning and servicing of plant and equipment can be carried out with ease, as operations involving difficult access or working conditions may not be undertaken effectively.

3.3　ACTUAL MAINTENANCE VISITS AND ACTIVITIES

From the questionnaires returned, it is apparent that the frequencies of maintenance visits vary considerably between operators and within individual organisations. Generally, such visits are made every two to four weeks although in one case daily visits are made.

Generally, routine maintenance work takes one hour per visit although in some cases up to four hours is necessary. Some stations, however, required prolonged maintenance visits primarily to desilt sumps or maintain old plant and equipment.

Pump manufacturers suggested that pumps only need an annual maintenance visit, although under their standard maintenance contracts they would normally visit twice a year.

The survey indicated the activities carried out at routine maintenance visits also varied considerably. In one case:

> "Pumping stations are visited every six weeks, or more often if required, to clean the sump, check that all valve handwheels are free to rotate, and carry out other checks as appropriate. M&E maintenance is carried out every six months."

Table 3.1 *Operational maintenance activities*[5]

Internal maintenance	Activity	Frequency
Pipework etc.	Clear blockage	As required
Wet wells	Clean walls and floor	As required
Electrodes or floats	Clean	As required
Lubrication systems – automatic	As manufacturers' recommendations	As recommended
Lubrication systems – manual	As manufacturers' recommendations	As recommended
Changeover duty pump – automatic	Remote control	As recommended
Changeover duty pump – manual	Operate changeover switch	Twice per year
Non-return valves	Check operation	Twice per year
Housekeeping	Clean windows, floor etc.	Quarterly
Hand-raked screens	Remove screenings	Weekly
Recording systems – automatic	Recover data	As required by database
Recording systems – manual	Recover data	As manufacturers' instructions
Standby generators	Run off load	Weekly
	Run on load	Monthly

Buildings and grounds	Activity	Frequency
Buildings, structures	Inspect and record condition (repair as necessary)	Annually
Fencing, gates etc.	Inspect and record condition (repair as necessary)	Weekly
Safety guarding on maintenance equipment	Inspect and record condition (repair as necessary)	Weekly
External and internal paintwork	Carry out painting	Every 3 to 7 years
Lawns	Cut grass	Twice in season
Paved areas	Control weeds	Once per year
Hedges	Trim	Once per year
Staircases, ladders, handrails etc.	Inspect and record condition (repair as necessary)	Once per year

Activities necessary for maintaining the buildings, structures and grounds etc. in sound condition, together with cleaning of wet wells, clearing blockages etc.

Table 3.2 Planned maintenance activities[5]

Equipment type	Activity	Frequency per year	Notes
Electrical equipment:			
Starters	Inspect, check	Twice	
Motors	operation, service,	Twice	
Transformers	clean and record	Once	
Circuit breakers	defects	Once	
Lighting, heating etc.		Once	
Standby generators		Twice	
Mechanical equipment			
Submersible pumps	Inspect, check	Once	
Centrifugal pumps	operation, service,	Once	
Compressors	clean and record	Twice	
Air vessels	defects	Once	
Progressive cavity pumps		Once	
Sump pumps		Once	
Macerators	Inspect, check	Once	
Mechanically raked	operation, service	Twice	
screens	and record defects	Twice	
Non-return valves		Once	
Vacuum pumps		Once	
Air release pipework and fittings		Twice	
Electronic equipment			
Telemetry systems	Service/inspect	Once	
Sonic level control transducers	Degrease	Twice	
Lifting gear			
Lifting beam and traveller	Inspect/test	Once	
			Statutory inspection of
Hoists	Inspect/test	Once	pressure vessels, lifting
Chains	Inspect/test	Once	beams etc. by insurance
Davits	Inspect/test	Once	company
Guide rails	Inspect/test	Once	
Fire extinguishers			
Fire extinguishers	Inspect	As manufacturers' instructions	

Activities necessary to ensure that the plant and equipment are maintained in good working order.

Another operator gave the following list of typical activities carried out during regular routine maintenance visits:

Check:

- for any damage caused by weather, vandals etc.
- the condition of covers, guide rails, chain hooks, kiosk and any ancillary equipment
- the sump for build-up of silt, rags grease etc. and if necessary, pump out
- the operation of floats/electrodes; clean if necessary
- the operation of lighting and emergency lighting
- the operation of telephones
- the operation of fire alarms
- the operation of standby generation plant.

Record:

- the flow rate at the discharge manhole (if possible) and determine whether it is normal
- the hours run (motors, standby generator etc.)
- the electricity meter reading
- the ammeter reading
- the time for the sump to empty to pump stop level (if possible)
- the weather conditions
- other data as required by in-house operational procedures.

Maintenance:

- clean/tidy the installation
- run/rotate standby and spare plant as recommended by the manufacturers
- annually paint kiosk.

Report:

- any defect identified in writing. If urgent, radio control centre, with written confirmation. If a major fault found, i.e. a choked pump, revisit station to check that repair has been carried out
- any regular build up of fat or silt.

In some cases, regular lubrication and minor adjustments to the pumping plant may also be required

Monitoring ammeter readings provides a good indication of the status and change of status of the pump when in operation.

3.4 SAFETY AND RELIABILITY

Strategic safety and reliability issues are addressed in a number of formalised procedures both at the design stage and during subsequent operation. These procedures include:

- health and safety plans at the various stages of design, construction/installation, operation and rehabilitation in accordance with the CDM Regulations[1]
- risk assessment
- HAZANS and HAZOPS
- zonal classification for hazardous areas
- confined-space entry
- permit-to-work systems
- electrical safety procedures
- efficiency condition monitoring
- availability and reliability studies.

3.5 COMMON PROBLEMS

The most common problems encountered in a routine maintenance procedure are rags, grit and grease, and, in some areas, vandalism. All can be reduced by careful design. BS 8005: Part 2[5] and CIRIA Report 121[4] provide advice on these matters.

3.5.1 Operational problems

The majority of operational problems encountered are due to defects in the hydraulic design of the pumping station, and this should always be given priority without neglecting the other aspects of design.

Rags tend to ball up and block pump suctions. Grit can reduce the capacity of the sump and, because organic solids tend to become trapped with the grit, septicity and odour may result. Grit may also cause increased wear of the impeller and casing. Grease will tend to form a scum on the liquid surface, build up on the walls of the sump and will be a source of odour as it decomposes. Problems with floating scum blankets appear to be increasing. In some situations they are becoming a serious operational problem due to adverse effects on the operation of level-control instruments, odour and access hazards.

One option would be to remove rags and grit before pumping, but this could create problems with odour nuisance, transportation and disposal. As there are often several pumping stations on a sewerage system, some of them being located out of sight in densely populated areas, it is often better to keep rags and grit in the sewage flow until it reaches the treatment works, where it can be removed. There are a number of automatic flushing systems available to clear blockages, keep rags and grit in suspension and break up scum. Thus it is recommended that rags, grit and scum should be retained in the sewage – the pumping station should be designed and the plant selected with this in mind.

3.5.2 Vandalism

It is necessary to design against vandalism. First, every effort must be made to prevent intruders from gaining access to the site. Additional protection should be provided to pumps, valve chambers and control panels, where damage can be caused to the plant and prevent it from operating satisfactorily. Intruders should also be prevented from gaining access to the sump, or any electrical equipment, where, in addition to interrupting the operation of the plant, they may cause serious harm to themselves or others. The following measures are recommended:

- secure perimeter fencing
- lockable covers should be provided to all underground chambers
- windows should be omitted from superstructures, and strong vandal-resistant doors should be provided
- it is virtually impossible to prevent paint being daubed onto walls etc., so consideration should be given using materials, or applying coatings, that facilitate the removal or obliteration of graffiti.

Underground stations are less prone to vandalism, but care should be taken in the location and choice of materials used for the control kiosk.

3.6 SUMP DESIGN

In order to prevent the build-up of grit and solids, pumping station sumps should be steeply benched, e.g. around 45°, and the horizontal floor area around the pump suctions kept as small as possible. Thus solid matter and grit should fall down the benchings and be drawn into the pump suctions.

The location of the incoming sewer needs to be considered carefully in order to avoid air entrainment in the pump suction. There are also some hydraulic problems that can be created by poor sump design, such as pre-swirl, air-entraining vortices, inappropriate sump velocities etc. These are all fully covered in *The hydraulic design of pump sumps and intakes*, by M J Prosser[12]. In some cases, hydraulic modelling of the sump may be appropriate.

To avoid cavitation in the pump, it is necessary to ensure an adequate net positive suction head (NPSH) over the operating range of the pump by maintaining a minimum sump level when the pumps are stopped. If the cut-out level in the sump is too high, it may encourage the build-up of grease and fat on the water surface. However, this can be countered by allowing the water level to fall low enough to create turbulence, breaking up fat and grease and drawing it into the pump suction. The extent to which this is feasible depends on the characteristics of the pump and should be discussed with the pump supplier to ensure that the overall performance of the pump is not impaired.

In larger pumping stations, where pumps cut-in in sequence, the rate and extent of deposition can be significantly reduced by incorporating cyclic alternation of the pumps, a variant of automatic duty pump rotation using the first to start, first to stop principle. In addition to continually shifting the suction points within the sump, significant reductions in sump storage volumes are possible. This in turn results in a lowering of operating depths and increased sump velocities. As a result sedimentation is reduced and with shallower pump sumps an opportunity is provided to reduce the static head. Alternatively, periodic drawdown of the sump to extra-low level by manual pump control may be beneficial.

3.7 PLANT SELECTION

Normally, pumps are specified with a design solids-handling capacity that will enable them to pass virtually anything that is likely to enter the sump. For typical sewerage applications, a diameter of 100 mm is usually quoted. However, where there is a significant risk of rags blocking the pump intakes, it has been common in the past to install screens at the inlet to the sump. It is now possible to fit an in-line macerator in the pump intake so that the screenings are retained in the flow, broken up and passed to the treatment works without causing pump blockages. This solution to the ragging problem can reduce routine maintenance operations and avoid disposal and other nuisances at the pumping station. However, due consideration should be given to the head loss through the macerator to ensure that adequate NPSH is available at the pump inlet and some means of inhibiting pump operation in the event of macerator failure.

Where a progressive cavity pump is to be used, care should be taken to keep flow velocities in the pump as low as practicable (which may produce operating speeds of significantly less than 1500 rpm), especially where grit is likely to be present, in order to reduce wear. This will have a cost implication at the construction stage, but there will be compensation in the form of savings in maintenance costs.

The build up of solids and scum in a sump can be controlled by pumped recirculation or aeration systems, designed to induce turbulence. An example of this, for submersible pumping stations, is a flushing valve, fitted to the pump casing. When the pump starts up, sewage is forced out of the flushing valve creating turbulence in the sump; after a few seconds, the valve closes and the sewage is forced up the rising main. In the case of dry-well pumps, auxiliary pipework and valves can be provided. Aeration systems should be used with great care as the introduction of air can lead to malfunctioning of the pumps.

Pressure washing facilities should be available. In a large pumping station, a suitable washwater booster set should be provided. For smaller stations, it might be preferable for the maintenance gangs to carry a small booster in their van, or on a trailer, which can be connected to the mains water supply at the station.

3.8 PUMPING MAIN DESIGN

Pipe diameter, flow velocity and operating head, together with material and cost, need to be considered together, in the design of a pumping main. The size and material selected should be the optimum based on an economic evaluation of whole-life costs.

Pumping mains should be designed to produce a self-cleansing velocity of not less than 0.75 m/s at minimum flows. However, the maximum velocity should not exceed approximately 2.4 m/s, otherwise the friction losses will be excessively high, increasing the head on the pump, and thus the pumping costs. Power costs are also addressed in Section 6.4.1. Depending on the pumping arrangement, the most economic velocities for the different numbers of pumps in operation at any one time should be determined.

In order to avoid blockages, the diameter of pumping mains should not normally be less than 100 mm. However, there may be situations, such as small developments in relatively remote locations, when a smaller pumping main is desirable. In such circumstances, a macerating pumping unit will almost certainly be required. If the use of a small main is also likely to generate high velocities and hence high heads, then a positive displacement pump (e.g. progressive cavity) should be considered.

As an illustration of the potential cost benefits of providing a small-bore pumping main, the relationship between the costs of providing and laying different sizes of MDPE pipe is given in Table 3.3.

Table 3.3 *Relative costs of MDPE pumping mains*

Nominal pipe size	Relative cost per metre
50 mm	1.0
63 mm	1.5
90 mm	2.2
125 mm	4.0
180 mm	8.9

Table 3.3 gives price differentials for one particular pipe material and is provided to indicate the potential savings, which must be set against the additional cost of a macerator. MDPE offers practical pipelaying benefits for small-bore pipelines that are not available with other pipe materials. It is probable, therefore, that, in considering pipeline materials, the designer would look at MDPE for a small-bore pumping main and a different material, e.g. DI or uPVC, for the larger main. It is important that the designer considers the relevant price differentials prevailing at the time he is carrying out his design.

Consideration needs to be given to carrying out a surge analysis of the pumping main system. The potential for positive and negative pressures due to surge is more likely to be significant in long mains or where velocities are high. Computer programs are available to assist with the surge analysis.

Surge pressures can be alleviated by various means, including:

- designed regulating or non-return valves in the main
- pressure regulation or surge vessels on the main
- flywheels on the pumps to avoid sudden shutdown
- a standpipe, close to the pumping station
- double-acting surge-relief valves on the system
- staged shutdown of pumps
- variable-speed drives.

It should be noted that the first method of surge alleviation is unsuitable for sewage containing solids. The last two methods cannot prevent surges in the event of power failure. This is frequently the most critical fault condition.

It is also necessary to consider the retention time of the sewage in the rising main, particularly if it is long, or where flows are intermittent. Each time the duty pump cuts in, a slug of sewage enters the main, and it progresses up the main in stages with each pumping cycle. At low flows, the retention time in the main can be excessive, resulting in septicity. This can lead to odour problems at the discharge point, and/or to acid corrosion in the downstream sewers. It is common to provide corrosion protection to the interior of the discharge manhole and the first few metres of the downstream sewer, or to use a corrosion-resistant pipe material. If odours are likely to be a problem, then measures should be taken to alleviate the septicity. A vent column should be provided at the discharge manhole to aid the dispersion of any gases released. Care should be taken at the rising main discharge to minimise turbulence that will encourage the release of any odorous gases and exacerbate any potential odour problem.

3.9 STATION PIPEWORK

CIRIA Report 121[4] indicates that, for wastewater applications, the simpler and more direct the pipework layout can be, the more successful will be its operation. Complications such as cross-connected pump suction pipework, which may appear to give operational flexibility, can significantly increase the risk of blockages resulting from strong swirling flows created by the T-connections. In some cases, the use of manifolded suctions can lead to air being drawn in through air-release valves on non-operating pumps, causing increased levels of vibration.

Provision of the following should aid effective operation and maintenance:

- simple, direct suction pipes to each pump from the suction bellmouth, of a suitable diameter, with a minimum of bends and other pipe fittings, but incorporating a suction isolation gate valve close to the pump
- delivery pipes connecting horizontally to a common manifold at high or low level, and then to a rising main or high-level sewer. (The reason for a horizontal connection is to prevent solid matter from dropping down the delivery pipes from pumps not currently operating.) If flow measurement is required, a magnetic-type flowmeter is generally most appropriate
- non-return valves in each delivery pipe. Isolating valves should be installed in such a position that the non-return valves can receive attention in situ when necessary. Valves should preferably be mounted in horizontal pipework
- pressure-relief valves on positive displacement pump deliveries, which should vent back to the sump
- air-release pipework on dry-well-mounted pumps, which should vent back to the sump
- pumped drainage in the dry well, discharging to the wet well at high level.

3.10 PROVISION OF ACCESS FOR ROUTINE MAINTENANCE OPERATIONS

Provision of safe access for all maintenance operations is the most important consideration in the design of component arrangements of a pumping station and its ancillary facilities.

Lifting arrangements, whether permanent or temporary, similarly need careful consideration and discussion with the future operators. There may be strategic policies in place covering these aspects. For example, there are statutory requirements for regular testing of all lifting equipment, so it may be more economical to use portable equipment, which is easier to maintain. Many operators are choosing this option.

The design should take account of the access and space requirements needed for the following:

- because every part of the pipework may be subject to blockage, the whole system must be accessible, or able to be dismantled, for cleaning. There should be access to each item of pipework and room to undo nuts and bolts and remove sections of pipe as necessary. Drain valves should be provided at the lowest points. In particular, flexible couplings should be fitted to the upstream sides of NRVs to permit easy removal for servicing or replacement
- lifting eyes should be provided on items of plant and pipework together with suitable access for lifting. Where possible, the need for double lifts should be avoided
- rodding points, such as hatch covers on pipe bends, should be considered and there must be room to insert the rods for cleaning. Although rodding is not common practice, it may be advisable in particular problem zones
- non-return valves are normally best positioned in a horizontal pipe close to the pump delivery branch before it joins the manifold. A bund wall or separate chamber around the valve is desirable, to collect spillage when dismantling and should drain into the wet well. Non-return valves frequently require dismantling to clear blockages (they are second only to pumps in this respect). The location of the non-return valve also requires careful attention so as to reduce friction at the spindle seals. Excessive friction can interfere with the proper functioning of the non-return valve
- the floor should be self-draining and non-slip to minimise the risk of slips, trips and falls in both wet and dry conditions.

In addition, fixed installation pumps should be provided with access into the pump (i.e. handholes with covers) to remove blockages. In some cases, pump casings need to be ventilated whenever the pump starts so suitable air-release pipework should be provided.

3.11 ELECTRICAL INSTALLATION

Factors having an impact on the operation and reliability of pumping stations to be considered at design stage include:

- the provision and sources of electrical supply
- the need for dual/standby electrical supplies
- access to switchgear and other electrical plant
- adequate sizing of switchgear and installation to take account of operational and fault conditions
- spare capacity and space for expansion/modification
- routing of cables
- comprehensive and clear labelling and instructions
- comprehensive indications of plant status and fault conditions. These would normally be displayed on the plant starter compartments, although a control station local to the plant could be considered
- adequate provision of power supplies for portable tools and equipment
- adequate lighting for operational and emergency conditions
- appropriate environmental control for plant operation (temperature, humidity, cleanliness of atmosphere, etc.)

- routine maintenance procedures
- choice of electricity supply tariff
- availability and storage of operation and maintenance information.

The regional electricity company (REC) will be able to advise on the reliability of the site power supply. Dependent on the criticality of the pumping station operation, backup power supplies may be required. These would normally be one of three types:

1. Dual supplies to the site. The REC may be able to provide feeders from separate primary substations, but this is not always possible or economic. The level of security is quite low, as it only gives protection against the loss of a single feeder or secondary substation, and maintenance/restoration of supplies is not within the control of the pumping station owner/operator, as it would be with a standby or portable generator.
2. A fixed standby generator. This has a high capital cost (for both the plant and civils structure) and considerable routine maintenance implications. Within the station design consideration needs to be given to the space requirements for the generator set, bulk and day fuel tanks and control panel. The provision of supply air for engine aspiration and cooling, hot air discharge and engine exhaust all need to be considered in detail.
3. Provision for connection of a portable generator set (normally a plug and socket connection for small loads or bolted connections for larger loads). Economically this may be the best solution. However, the overall logistics of portable generator provision for the region and the time taken in delivering and connecting the portable generator have to be taken into account.

On-site reliability can be improved by arranging the switchboards, motor-control centres and electrical distribution into two sections. Duty and standby plant should be connected to separate sections.

The location of hand-operated controls, emergency stop facilities and isolation points will require consideration. Hand-operated controls may be located adjacent to the plant, marshalled at a common location and/or located on the motor-control centre. Emergency stop facilities will normally be located adjacent to individual items of plant. Dependent on the plant layout, access etc., consideration should be given to emergency stop facilities controlling groups of plant. Isolation will normally be provided at each motor starter and may also be provided adjacent to plant. External equipment should be arranged to prevent unauthorised operation or vandalism.

3.12 AUTOMATION

The following activities should be carried out automatically:

- pumps switched on and off by level control
- rotating duty and standby pumps – autosequencing
- automatic lubrication systems
- restart after power failure, or tripping out due to overload or loss of prime
- timer for starting ancillary equipment such as macerators, sump pumps, flushing valves, agitators, heating elements etc.

Where installed, the following can significantly reduce the frequency of visits by the maintenance gang:

- automatic backflushing facilities to clear blockages
- automatic septicity monitoring and dosing
- automatic ventilation of confined spaces
- automatic wet-well cleaning.

3.13 INSTRUMENTATION, CONTROL AND AUTOMATION (ICA)

Sewers for adoption (4th edition)[11] recommends that the following operational states should be monitored remotely by means of telemetry:

- emergency overflow operating
- high-level wet well (from float switch) and back-up float control operating
- mains power failure
- level control system failed
- No 1 pump tripped (starter)
- No 2 pump tripped (starter), and so on depending on the number of pumps
- No 1 pump running
- No 2 pump running, and so on depending on the number of pumps.

The following additional items may also be provided:

- intruder alarms
- sump level
- hours run
- low wet-well level.

The information obtained may be used simply for monitoring the operation of the pumping station, responding to emergencies and initiating maintenance visits, or it may be used as part of a sophisticated regional control system.

Control systems need to be:

- self-contained
- simple
- robust
- reliable (including back-up facilities if necessary)
- appropriate for their physical and environmental location.

During the design of communications and telemetry systems, address:

- level-control systems
- high- and low-water-level alarms
- off-site alarms (type and status)
- off-site control (including upstream and downstream facilities)
- off-site supervision
- medium to be used (e.g. telephone, radio, mains-borne signalling, etc.)
- reliability (including back-up/standby equipment and power supplies, e.g. battery backup, if necessary).

The general rule when designing the controls for pumping plant must be to keep it as simple as possible given the requirements of the pumping process. There is no benefit from using complex and expensive processor-based controls when control could be obtained from discrete switching points in a simple level controller. For small packaged pumping stations the use of manufacturer's standard control panels should be considered, although these may be restricted by the process control requirements and this type of panel's limited signalling facilities.

The provision of basic instrumentation such as motor current ammeters and hours-run meters allow a reliable history of the pump to be established. Any unexplained changes from the norm will be an indication of some pending maintenance problem, such as wear, ragging up, etc.

Flow and level detection devices should preferably be of the non-contact type. Float switches or electrodes could be used satisfactorily for monitoring high-level alarm conditions where they are not normally in contact with the sewage.

To prevent spurious control signals, plant operation should be verified by positive feedback (e.g. pump running should be detected by monitoring flow and not that the motor starter main contactor is closed).

Pump/plant monitoring of, for example, high temperature, seal leakage, loss of rotation, should be hard-wired into the motor starter. It is not recommended to take these safeguard facilities through indirect control such as programmable logic controller (plc) or distributed control.

There is a growing trend towards transmitting substantial amounts of complex data from site, i.e. all status and fault indications, for off-site analysis. In practice this is costly (both in terms of capital and revenue expenditure) and often results in a vast amount of data being processed and stored off-site which is generally not accessible by, and of little use to, those responsible for the day-to-day operation and maintenance of the plant. Consideration should be given to keeping the information required for off-site monitoring to a minimum.

Four levels of complexity for off-site monitoring could be considered. The level adopted is likely to be subject to established local policy requirement, which will dictate the actual signals to be provided.

1 *Level 1 No off-site communication*

This may consist of a common signal representing a combination of various alarms, the result of the failure. This will not usually be acceptable where a pollution incident could result.

2 *Level 2 Alarm only*

This may consist of a common signal representing a combination of various alarms, e.g.:

> "hihi" level alarm in wet well
> plant failure
> intruder alarm, fire alarm etc.

The use of this level of monitoring will be dependent on the speed of response to the alarm and the likely result of any failure.

3 Level 3 Alarms and limited-status signals

Typically this may consist of:
    "hihi" level alarm in wet well
    plant RUNNING/STOPPED
    plant AVAILABLE/NOT AVAILABLE for AUTO control
    plant HEALTHY/FAULT
    fire alarm HEALTHY/FAULT
    fire alarm HEALTHY/ALARM
    intruder alarm HEALTHY/FAULT
    intruder alarm HEALTHY/ALARM
    standby power plant RUNNING
    telemetry system/comms path HEALTHY/FAILURE.

4 Level 4 Comprehensive alarms, faults and status indications

Typically these may include:

"hihi" level alarm in wet well
wet well level (analogue)
level-detection equipment HEALTHY/FAULT
hi level (assist plant START switching point)
int level (plant START switching point)
low level (plant STOP)
plant RUNNING/STOPPED
plant AVAILABLE/NOT AVAILABLE for AUTO control
plant in AUTO, HAND or OFF
plant HEALTHY/FAULTS (overload, seal leakage, over temperature, flow failed, etc.)
plant motor absorbed current (analogue)
power HEALTHY/FAILURE
standby power plant RUNNING/STOPPED
standby power plant alarms (e.g. low oil pressure, over temperature, frequency deviation, voltage deviation, etc.)
telemetry system HEALTHY/FAILURE
communication path HEALTHY/FAILURE
fire alarm HEALTHY/FAULT
fire alarm HEALTHY/ALARM
intruder alarm HEALTHY/FAULT
intruder alarm HEALTHY/ALARM etc.

Clearly the greater the number of signals required off-site the greater the capital and revenue costs for on-site and off-site equipment.

3.14 HAZARDOUS AREAS

All areas should be subjected to a hazardous areas assessment. It should be noted that a hazardous area classification is determined by the probability of an explosive atmosphere existing and should be undertaken regardless of whether or not electrical and ICA plant is installed in the area concerned. The code of practice for the selection, installation and maintenance of electrical apparatus for use in potentially explosive atmospheres is to be found in BS 5345. This defines a "hazardous area" as one "in which explosive atmospheres are, or may be expected to be present in quantities such as to require special precautions for the construction and use of electrical appliances". The electrical and control equipment is chosen to suit the area classification.

Electrical and ICA plant installed in hazardous areas should be kept to a minimum due to its high cost. Such equipment is subject to special considerations, e.g. special enclosures, protection by intrinsically safe barriers, isolation of neutral conductors, etc.

It should also be noted that some "soft" metals (e.g. aluminium, titanium etc.) may create sparks when in contact with ferrous metals. Consideration should be given to the proximity of such metals in hazardous areas.

Portable electric tools must be suitably classified for use in a hazardous area.

Adequate warning notices must be provided advising of hazards and designating the appropriate confined space category.

3.15 COMMUNICATION

Those responsible for the design and construction of the pumping station (including the manufacture, supply and installation of plant) should produce operating and maintenance manuals and maintenance schedules. It is essential that these documents are well set out, logical and easy to understand so that safe and efficient operating and maintenance procedures can be established.

Under the CDM Regulations[1] the planning supervisor must ensure that a health and safety file is prepared, which should be handed over to the owner on completion of construction and commission of the works. The health and safety file should include the O&M manual, which should contain details of all the health and safety provisions, and recommended procedures, relevant to the safe operation and maintenance of the pumping station, such as entry procedures and electrical safety regulations applicable to the operating authority.

3.16 CASE HISTORIES

Where operators have given details of pumping stations that require frequent or prolonged maintenance visits, frequent emergency call-outs, or that require very little maintenance, there is little or no indication of the reason for the problem. Where the reason is identifiable, details are given in the following sections.

3.16.1 Frequent emergency call-outs

Case Study 1

- operation of the pumps is duty/standby, with duty changeover in sequence
- No 1 Duty Pump runs and fails
- Standby (No 2 Pump) starts and pumps through its cycle
- sequential operation now calls for No.1 Pump to operate, but this is in failed condition, so neither pump operates
- high-level alarm.

Clearly, the fault is in the design of the pump control, which should ensure that if one pump has failed, the healthy pump always operates.

Case Study 2

In a second case relating to frequent call-outs, the fault has been attributed to two factors: pumps that were susceptible to wear, and badly adjusted low-current trips.

Other reports of frequent emergency call-outs are associated with high flows during rainfall and the inability of the pumping plant to cope. Old and unreliable pumps were also the cause of frequent emergency call-outs.

Case Study 3

This was originally an ejector station with an inadequate output, which suffered from frequent flooding and overflows, together with odour problems. In the mid-1970s it was replaced with a submersible pumping station, which suffers from problems with siltation and build up of fats, requiring frequent maintenance visits. Potential solutions considered by the operator are a run-on feature on the pump control and/or an automatic flushing valve.

3.16.2 Trouble-free operation with low levels of maintenance

Several pumping stations were listed under the above heading, but it was not possible to discern any particular features that contributed to this.

3.17 CHECKLIST

This list, in the form of actions, sets out key issues to be addressed in establishing designs for ease, economy and reliability of routine operation and maintenance.

Action	Reason	Section ref.	Checked	
General access				
Ensure a permanent vehicular access to the site	Site layout	3.2	/	/
Provide adequate parking and turning space	Site layout	3.2	/	/
Allow room for a vehicle-mounted washwater booster, a mobile pump or a generator to stand close to the pumping station	Sump cleansing Emergency power	3.10 3.11	/	/
Plant design				
Ensure that the plant is suitable for the required duty and capable of passing solids of an appropriate size	Reliability	3.4	/	/
Check that the pipework is designed to minimise the risk of blockages and to ease clearing any blockages that might occur	Operational problems	3.5.1	/	/
Check that the plant is suitable for any relevant hazardous area classifications	Safety	3.14	/	/
Sump design				
Check the sump capacity is suited to the application and pump starting limitations	Hydraulic problems	3.6	/	/
Ensure the sump design is the best possible to avoid hydraulic problems with the pumps	Reliability	3.6	/	/
Minimise air entrainment	Prevent cavitation	3.6	/	/
Evaluate the sump design with regard to its efficiency in minimising the deposition and build-up of grit and other solids	Reliability	3.6	/	/
Apply appropriate hazardous area classifications	Safety	3.14	/	/
Facilities for cleaning				
Provide adequate access to the sump for cleaning, with appropriate numbers and sizes of covers, access ladders/stairways	Maintenance safety	3.2 3.10	/	/
Provide adequate ventilation to enable cleaning operations to be carried out easily and safely	Maintenance safety	3.2 3.10	/	/
Consider whether a pressure washwater system is required for efficient cleaning of the sump	Maintenance	3.2 3.10	/	/
Determine whether a booster unit should be installed at the station or carried on the maintenance vehicle	Plant selection	3.7	/	/

Action	Reason	Section ref.	Checked	
Consider the most appropriate way of safeguarding the power supply, as well as the noise and exhaust gas emission implications of any standby power generation required	Services	2.8	/	/
Consider fitting automatic flushing valves to pump volutes	Plant selection	3.5.1 3.7	/	/
Ensure the pipework arrangement facilitates cleaning/unblocking of delivery pipework, if required	Layout	3.2 3.9	/	/
Access to plant				
Ensure adequate access to all plant and equipment	Safety	3.10	/	/
Provide access openings of sufficient size to facilitate the installation and removal of plant and equipment	Maintenance	3.10	/	/
Determine the kind of lifting equipment to be provided and its suitability for the pumping station.	Maintenance	3.10	/	/
If portable lifting plant or equipment is to be used, provide adequate access	Maintenance	3.10	/	/
Communication facilities				
Determine the appropriate level of telemetry for the pumping station	Telemetry	3.13	/	/
Establish what emergency alarm facilities are required	Telemetry	3.13	/	/
Decide whether signals are to be communicated by telephone, dedicated landline or radio. If by radio, establish whether there are any constraints on the type, size and height of the aerial	Telemetry	3.13	/	/
Rationalisation				
Decide on the minimum level of complexity of plant and control equipment necessary for the satisfactory reliable operation of the plant and consider whether there is any justification for exceeding this	Plant selection	3.7	/	/
Consider whether maintenance operations can be made more efficient by promoting uniformity/ standardisation of plant and equipment within the operating district	Plant selection	3.7	/	/
Health and safety				
Comply with the CDM Regulations[1] and prepare a health and safety file	Statutory requirement	3.4 3.15	/	/
Carry out a hazardous areas analysis and ensure that plant is suitably rated	Safety/reliability	3.4 3.14	/	/

Action	Reason	Section ref.	Checked	
Carry out a hazards analysis and hazardous operations study (HAZANS and HAZOPS) and take account of the recommendations in the design	Safety/reliability	3.4 3.14	/	/
Provide adequate warning notices at the station regarding hazards and confined spaces	Safety	3.14	/	/
Decide whether it is appropriate to provide sanitary, washing and/or messing facilities	Amenities	3.2	/	/
Vandalism				
Evaluate the level of risk of vandalism at this site	Security	3.5.2	/	/
Determine which measures would be appropriate to minimise damage in respect of site security, lighting, security cameras, types and design of superstructure (avoidance of windows) and construction materials (damage-resistance and/or ease of cleaning or overpainting)	Security	3.5.2 3.13	/	/
Manuals				
Provide a comprehensive operation and maintenance manual that sets out full procedures for the routine maintenance of the station including relevant health and safety requirements	Communication	3.15	/	/
Check that the manual offers sufficient information in the right form to enable the work to be carried out	Communication	3.15	/	/
Check that the manual highlights the need for regular certification of lifting devices	Safety	3.15	/	/
Decide where the manual is to be kept	Communication	3.15	/	/
Ensure that the station maintenance staff have ready access to the O&M manual	Communication/Safety	3.15	/	/

4 Major repairs and maintenance

4.1 INTRODUCTION

Routine operational maintenance is covered in Section 3. This section covers relatively major maintenance operations such as the repair or replacement of pump and motor components that are worn, burnt out or have otherwise reached the end of their useful lives. Such operations should only need to be carried out occasionally. Sometimes major maintenance is required following damage caused by foreign objects in the sewer, such as builder's rubble containing bricks, stones, sand and grit, which can cause increased wear to the pump casing and damage to the impeller, and can stop the pump and stall the motor. So far as possible, such matter, which can also cause blockages to the upstream sewer, should be excluded from the system.

4.2 COMMON PROBLEMS

The most common problems encountered when carrying out major maintenance and repairs concern access, ventilation and lifting facilities. These may occur for the following reasons:

- the designer does not adequately take account of the maintenance needs of the plant and the potential hazards of working in a confined space
- the necessary facilities were not provided, in order to contain costs at construction stage, as major maintenance was not envisaged for several years.

The first point above can be overcome by effective communication between the designer and the pump supplier at an early stage during the design of the pumping station. Compliance with the CDM Regulations 1994[1] should encourage better communication in the future. The second point relates to capital expenditure policy at the time of construction, but savings in capital cost may turn out to be false economy. This aspect is covered more fully in Section 6.

4.3 ACCESS, WORKING SPACE AND VENTILATION

With careful design, it should be possible to provide good access and the necessary working space without providing an inordinately large building. For example, some progressive cavity pumps have significant space requirements for stator withdrawal. This need not be a problem if the main plant access door is located at the delivery end of the pumps, as the shaft can be withdrawn through the door if necessary.

As mentioned in Section 2.5, there can be a problem with deep submersible pumping stations, where the valve chamber is located at nominal depth adjacent to the sump. If the distance between the bottom water level and the non-return valve is greater than about 8 m, it will be difficult to ensure proper operation of the pumps. A cost-effective solution would be to locate the valves on the vertical delivery pipes within the sump, together with suitable access facilities. The preferred solution would be to provide a wet-well/dry-well pumping station, with the valves mounted horizontally at low level. In some circumstances, suitable air valves can be installed in the valve chamber adjacent to the non-return valves. However, this should be done in consultation with the pump manufacturer.

All pumping stations should be ventilated, both to help prevent the build-up of hazardous gases and to promote good ventilation of the sewerage system. Additional ventilation may be required to ensure the safety of workers, and appropriate provisions should be made. Typically these requirements will entail venting of the wet well and forced ventilation of the dry well.

4.4 COMMUNICATION

A dialogue with the pump manufacturer about maintenance requirements should be carried out as part of plant selection. Concerns about potential problems need to be discussed. If a manufacturer is unable to provide a satisfactory answer, he should be excluded from the tender list, or where appropriate, suitable design modifications should be agreed and incorporated into the specification. If a pumping station designer fails to consult fully with the manufacturer(s), he is in danger of producing a specification that the manufacturer considers to be unreasonable or unrealistic, and this will be reflected in his price. In the event of subsequent problems, there may be some difficulties in determining where the responsibility lies.

In one case, quoted by a manufacturer, the designer had specified measures to provide an alternative cooling system, instead of the standard integral pumped media cooling system, in order to eliminate the risk of cooling jacket blockages. The manufacturer considered that the auxiliary clean-water flushing system specified was completely unnecessary and would have added additional actuated controls, increasing the number of system components, failure of which would cause the pump to be shut down. The design and operation of the pumps in question had been proven over many years, handling unscreened sewage. The design relied upon a small volume of the pumped media for cooling, which was effectively screened by the impeller upper shroud and pump casing running clearance. No failures of sewage pump units resulting from blocked cooling jackets could be identified by the manufacturer.

4.5 MAINTAINING FLOWS DURING MAINTENANCE OPERATIONS

It is essential to provide adequate means for maintaining flows if any maintenance operations are likely to take longer than a couple of hours.

For small pumping stations this means the provision of facilities for overpumping by portable pump. A suction pipe may be provided from the sump to the valve chamber, terminated with a valve and a quick-release coupling; a valved tee is also needed on the rising main inside the valve chamber, fitted with a quick-release coupling, as illustrated in Figure 2.2. Alternatively, a flexible suction may be dropped into the sump. If this solution is chosen special flaps should be provided, as otherwise the cover will have to remain open at night and suitable protection provided.

On larger pumping stations, a divided sump may be provided, with a penstock in the division wall to balance flows during normal operation. This enables half the sump to be isolated for maintenance, the remaining pumps dealing with the flows. An overflow should be provided to handle excess flows in the event of a storm. Provision can be made for sewage in one half of the sump to weir over the division wall if an emergency should occur during maintenance, such as a heavy storm or the failure of the duty pump. Arrangements should be made to ensure that the high level alarm sounds early enough to enable anyone working in the sump to evacuate it.

Each pump delivery should have an isolating valve downstream of the non-return valve, to enable the non-return valve and pump to be dismantled for maintenance. In wet-well/dry-well pumping stations, an isolating valve should also be provided on the suction pipe to each pump. These valve arrangements are illustrated in Figures 2.2 and 2.4.

In case of a power supply failure during maintenance operations, provision should be made for an alternative power supply, as discussed in Section 3.11. Alternatively, for a small pumping station that is provided with connections for overpumping, a self-contained portable pump with its own power source might be used. Apart from the incoming distribution board, all the electrical plant and equipment should be capable of being isolated for maintenance without shutting down more than one pump at a time. This will generally preclude the use of wardrobe-style switchgear.

4.6 ELECTRICAL AND CONTROL SYSTEMS

Factors having an impact on the maintenance of pumping stations that should be considered at the design stage include:

- accessibility of switchgear and other electrical plant
- accessibility of components within switchgear etc.
- adequate labelling of components (ratings, references, weights where appropriate, etc.)
- cable routes
- the need for alternative power and switching arrangements (possibly with manual control) during maintenance
- minimising the amount of equipment in hazardous areas and confined spaces
- suitable and accessible isolation for equipment mounted in hazardous areas and confined spaces
- adequate lighting for operational and emergency conditions
- adequate provision of power supplies for portable tools and equipment.

4.7 MAINTENANCE CONTRACTS

Traditionally, maintenance work has been carried out by the owner's in-house team of maintenance engineers, or by agency authorities. Some water service companies are phasing out agency agreements so that they can have more direct control of sewerage operations. Often, the old agency wins a contract with the water service company to carry out routine maintenance work, as before, except that the water service company provides closer supervision of the work. Usually, these contracts cover the regular "monthly" visits (or as required) to clean out the sump and carry out routine inspections. They may also include the half-yearly inspections of the mechanical and electrical plant and equipment, although, if major repair work is required, the manufacturer or another competent M&E engineer will often need to be called out to carry out the necessary repairs.

Increasingly, M&E maintenance is being let on an open-tender basis. Many of the pump manufacturers offer various types of maintenance contract covering not only their own plant, but also other manufacturers' plant. Typically, for an annual fee, the contract includes an annual (or twice-yearly) inspection with a report and an estimate for any work that is recommended. For an enhanced annual fee, the replacement (as required) of wearing parts can be included, or the contract might include all parts and labour necessary to keep the station in optimum operating condition.

Before entering into any contract that includes more than the basic inspection and report, the maintenance contractor would normally inspect the plant to determine its condition and the level of risk he is taking on. Some initial repair/maintenance work may have to be carried out, at the expense of the owner, before the contractor will accept a contract. In some cases, e.g. for old plant, the contractor will only accept a contract to undertake the basic inspection and report, all additional work being at the owner's expense.

The main advantage of the enhanced maintenance contracts is that some or all of the risk is borne by the contractor, and the owner can budget more accurately for his maintenance costs.

4.8 SIMPLICITY AND STANDARDISATION

Traditionally, many pumping stations have been constructed using competitive tendering to select the contractor. There are many examples where pumps in different pumping stations in the same district have come from a large number of sources. Since most of these pump manufacturers also go to tender for their control panels, the variety of control panels in the district is even wider than for the pumps. Furthermore, many of these control panels are complex, often incorporating instrumentation, control and telemetry equipment.

This lack of uniformity can create problems with spares availability and the ability of maintenance staff to carry out even simple operations without reference to the drawings and/or a manual, which may not be readily available.

Most maintenance gangs consist of operatives who are capable and versatile, but who are not highly skilled technicians. They can normally handle routine M&E maintenance operations and deal with emergencies, in addition to cleaning out the sump and other routine matters. Designers and managers – many of whom have a propensity for high-tech solutions – should be aware of this.

Thus, two important aspects of designing for ease of maintenance are as follows:

1. Keep it simple
2. Standardise.

It has been suggested that operatives would rather see a float that goes up and down than an ultrasonic head that apparently does nothing. During consultations, an example was given of a pumping station that started pumping continuously immediately after a maintenance visit. There was no obvious reason for this until eventually it was found that the ultrasonic level detector in the sump had been turned round accidentally by an operative when cleaning the sump. This may be a little extreme, but the avoidance of complex control systems and telemetry will provide the following:

- greater scope for the operative to carry out necessary repairs, instead of calling out a specialist to deal with it
- savings in cost (if a maintenance gang visits the station frequently to carry out routine maintenance, as described in Section 3.2, the checks that are made during such visits should be sufficient to eliminate the need for expensive fault monitoring and detecting telemetry systems).

Standardisation offers many advantages, which are summarised below:

- availability of spares: a smaller range of spares will be required making it more economic to keep a local stock of spares instead of ordering them from the manufacturer when required. It may be practicable to carry some of the most frequently used spares in the service van
- interchangeability: in an emergency, it will be easier to keep stations in operation if there is a reasonable level of interchangeability of pumps and some parts, even on a temporary basis
- staff training: staff can be trained to be familiar with a relatively few designs and systems, instead of trying to cope with many
- use of standard manuals and drawings: all too often, wiring diagrams and other items of literature are lost and duplicates cannot be found because the equipment is a "one-off". Standardisation would reduce the number of drawings required, enabling operatives to carry copies in their vehicles, and for duplicates to be held securely back at base.

4.9 CASE HISTORIES

4.9.1 Problems encountered during maintenance work

Case Study 4 – Pumping station located at golf links:

- problems: no lifting gear available
  no hardstanding for mobile crane
  damage to access road
  portable pump used for maintaining flows blocked frequently, required daily maintenance and was difficult to refuel
- safety: the security fencing (palisade type) did not provide adequate protection from flying golf balls.

4.9.2 Pumping stations where maintenance was particularly easy

Case Study 5 – Submersible pumping station, where motors had burnt out:

- lifting equipment: portable A-frame with block and tackle
- particular feature: clear access directly over pumps with space to arrange lifting equipment.

Case Study 6 – Foul and surface water pumping station with wet well/dry well and brick superstructure:

- lifting equipment: overhead crane in pumping station
- particular feature: generous space allowances; circular chamber with inner dry well and outer wet well with pumps well spread out.

Other examples where pumping station maintenance was particularly easy were given in response to the questionnaires. The main reason was adequate provision of both access and lifting equipment. Where a superstructure was provided the key feature was adequate lifting equipment; where there was no superstructure, the important features were large removable covers giving good access to the pumps from overhead, enabling a tripod and block and tackle to be used.

4.10 CHECKLIST

The following checklist, in the form of actions, sets out the key issues to be addressed in establishing designs for ease and convenience of other maintenance.

Action	Reason	Section ref.	Checked	
Equipment to be provided				
Decide whether the major maintenance will be carried out by in-house engineers, a general maintenance contractor (with specialists employed when required), a specialist M&E contractor, or the pump manufacturer	Operating costs	4.7	/	/
Specify what equipment is to be carried by the maintenance engineers and what should be provided at the station. This relates particularly to lifting equipment	Maintenance operations	4.7	/	/
Establish what loads the station lifting equipment will be need to lift	Safety	4.7	/	/
Access, working space and ventilation				
Decide whether pumps will be removed for inspection and repair or if work will be carried out in-situ	Consider hazards	4.3	/	/
Ensure that there is sufficient working space for the work to be carried out as required	Layout	4.3	/	/
Consider what additional ventilation will be required while people are working at the station and how this will affect the design	Safety	4.3	/	/
Communication				
Consult the pump manufacturer about measures for achieving optimum reliability by reducing the frequency for "major" maintenance	Reliability	4.4 4.5	/	/
Consult the pump manufacturer about lifting requirements for maintenance of the plant	Plant selection	4.2	/	/
Consult the pump manufacturer about access, working space and ventilation requirements	Layout	4.2 4.4	/	/

Action	Reason	Section ref.	Checked	
Maintenance of flows				
Decide on the most appropriate method of isolating pumps and maintaining flows during maintenance, e.g. by: isolating valves on suction and delivery pipework divided sump with emergency facilities provision of facilities for over pumping standby power supplies	Reliability/maintaining operation	4.5	/	/
Rationalisation/standardisation				
Investigate the scope for standardisation to improve spares availability, interchangeability of parts or whole units, speed and efficiency of maintenance work	Whole-life cost	4.8	/	/
Health and safety				
Establish which safety provisions will be required, e.g. emergency cut-outs, guard rails etc	Control/safety	4.6 4.8	/	/
Manuals				
Ensure that the O&M manual provides sufficient information in the right form to enable the work to be carried out	Training/standardisation	4.8	/	/
Check that all special provisions and procedures for health and safety during major maintenance are set out clearly in the O&M manual. Make sure that these provisions are identified in the health and safety file	Safety	4.8	/	/

5 Rehabilitation

5.1 INTRODUCTION

Design lives of civil works tend to be about 60 years, whereas the design lives of plant and equipment are normally around 20 years. Thus, the structure may still have many years of useful life remaining after the plant is beyond economic repair.

Rehabilitation involves the replacement or enlargement of plant utilising as much of the original building as possible, with a minimum of structural alterations.

There are several reasons why a pumping station may need rehabilitation:

- plant is beyond economic repair
- plant needs upgrading to cope with increased flows
- there is a need to utilise new technology
- there is a need to improve efficiency and reduce O&M costs.

5.2 REPLACEMENT OF WORN-OUT PLANT

The simplest scenario is to replace the worn-out plant with plant that is identical with the original. However, after 20 years or more, it is likely that the model will have been superseded. In replacing the plant it is generally desirable to incorporate recent developments in pumping technology. This may require modifications to the structure. The aim, in designing for ease of rehabilitation, is to ensure that any structural modifications, and the costs thereof, are kept to a minimum without compromising contemporary health and safety requirements.

5.3 UPGRADING EXISTING PLANT

It is likely that the required duty of the pumping station will have changed during the years that it has been in existence. Flows may have increased and/or the pumping head changed. The latter may be due to increased flows in the existing pumping main, or because of changes in the sewerage system downstream of the pumping station. This will normally necessitate larger pumps.

It is common these days to design for reasonably short design horizons, and the design life of the plant of approximately 20 years is probably the longest horizon it is practicable to design to. However, it is worthwhile giving some consideration to what may happen at the end of that time. Is it likely that there will be a sustained growth in the population served beyond the design horizon, or will it increase to around the design population and then remain static? How does this pumping station fit into the overall sewerage strategy? Will it increase in importance, or is it likely to become redundant? While much of this is speculative, it should be considered, particularly if whole-life costs are used as a means of project evaluation at the design stage.

5.4 UTILISATION OF NEW TECHNOLOGY

In recent years there have been significant developments, such as improved pump materials, more reliable mechanical seals, high-efficiency motors and improved control and automation systems. It is likely that this progress will continue into the future as moves are made towards more centralised control and monitoring of pump performance with a view to minimising visits by maintenance gangs (see Section 4.8). Thus it is likely that there will be a need to allow space to enable new and more up-to-date equipment to be provided.

5.5 IMPROVED EFFICIENCY AND REDUCED OPERATION AND MAINTENANCE COSTS

With the current emphasis on reducing the use of energy so far as possible, research is constantly being carried out on the design of increasingly more efficient plant, and it is likely that such plant will need to be incorporated during any pumping station rehabilitation.

If the current trend of reducing workforce costs continues to be operationally acceptable, then future developments are likely to include new technologies:

- for the automatic cleaning of pump sumps
- for preventing the build-up of grease/fat
- for reducing the deposition of grit
- for preventing the blockage of pumps and suctions by rags etc.

Such technologies are already in existence, but there is scope for further development. Designs should take account of the implications relating to space requirements if the introduction of such technology is envisaged.

5.6 DESIGN OPTIONS

When designing for ease, speed and economy of rehabilitation, it is necessary to consider what will be the most likely reasons for that rehabilitation. In considering the various possible design options, it is important to take an intelligent look at whole-life costs and to compare realistic alternatives. It is all too easy to take the life of the pumping station as the 60-year life of the civils works and to assume that the plant is replaced with exactly the same plant after 20 and 40 years to determine the appropriate net present value for comparison purposes. However, as already demonstrated, this is unlikely to be realistic. There must be an attempt to make a realistic prediction of future rehabilitation needs, in order to carry out a sensible analysis, and this should be taken into account in costing and designing the civil works.

It may be that the whole pumping station will become redundant when the plant reaches the end of its useful life and, if possible, this should be taken into account at the initial design stage.

There may be several alternatives under consideration for future development, and as a consequence, making it impossible for the designer to come to a rational prediction about future needs. In this case, he must decide whether it is sensible to assume a design life for the civil works that is greater than that of the plant it contains. In other words, will it be practicable to rehabilitate the pumping station structure when the plant reaches the end of its useful life, or is it likely that it will have to be demolished and replaced?

There are thus two options:

1. design a structure with a short (e.g. 20-year) design life that can be easily replaced when the time comes, or
2. design a structure with a long (e.g. 60-year) design life, but build in as much flexibility as possible to facilitate future rehabilitation.

The option chosen will have an important influence on the selection of materials and methods of construction, and on the initial cost.

If flow increases during the early years of the pumping station's life are predicted, consideration should be given to increasing the pumping capacity by designing for changes in impeller. The percentage increases that can be achieved will be specific to each individual pump's operating curve. Pump casing and motor sizes may limit the scope of this option.

5.7 CURRENT TRENDS

There are many pumping stations that are likely to need rehabilitation in the foreseeable future but which may not be suitable for the installation of new or upgraded plant. As a result, major modifications will be required or an otherwise valuable asset may have to be demolished. However, with the development of submersible pumps, the trend is for pumping stations that are constructed almost entirely below ground, apart from the electrical control gear and telemetry, which is normally housed in a specially designed prefabricated cubicle.

Where a superstructure is provided, there is a tendency to move away from the traditional bricks and mortar in favour of modern lightweight materials such as GRP and profiled steel. The reason for this is usually cost, together with ease and speed of construction. However, these materials have relatively short design lives (figures of 15 to 20 years are sometimes quoted) and would normally need to be replaced more often.

Thus, the most permanent part of the structure will be the part that is below the ground. This comprises the wet well and valve chamber in a submersible pumping station, the sump and pump chamber in a wet-well/dry-well pumping station and the sump only in an above-sewage pumping station.

5.8 BELOW-GROUND STRUCTURE

The feature common to all the above types of pumping station is the wet well or sump. The two characteristics of the sump that should be taken into account when designing for ease of rehabilitation are as follows:

1. capacity
2. shape.

The capacity of the sump is based on the required number of pump starts per hour and the control philosophy used. Pump manufacturers usually state the maximum permissible number of starts per hour for their pumps. If this figure is used for design, it will give the most economic size of sump for the installed pumps. However, if the sump is designed initially for a reasonably low number of pump starts per hour, this will enable larger pumps to be installed later without any change to the sump size. An element of care is required to ensure that the frequency of starting is not too low in order to avoid septicity and sediment deposits in the sump. Provided self-cleansing velocities are achieved during pumping, deposits will be of a transitory nature.

It may be necessary to change the shape of the sump to accommodate larger pumps. Benching should not be constructed integrally with the concrete walls, or in structural strength concrete, as it will be very difficult to remove. However, if it is constructed in a weak concrete, sealed with a dense rendering, it will be much easier to break out and re-form in the future.

If it is necessary to increase the capacity of the sump, this can be achieved by building a new sump alongside, interlinked with the existing one. If the pumps are then distributed between the two sumps, this arrangement will give all the benefits of a divided sump and avoid the need to demolish the old sump. This is particularly applicable to submersible pumping stations; however, it presupposes that there is adequate space on the site.

5.9 LAND ACQUISITION

There is a temptation to save on land costs by having a compact site, especially if temporary access can be provided over adjacent land during construction. This should be resisted. Encroachment of other buildings towards the site boundary may make it almost impossible to carry out rehabilitation work because of the lack of suitable access, working space and hardstanding for temporary plant or equipment, both during construction and operation. The probable rehabilitation needs should be considered together with the construction requirements when deciding the area of land to be acquired.

5.10 SUPERSTRUCTURE

If the pumping station cannot be accommodated below ground and a substantial superstructure is required that is likely to outlive the plant, as much flexibility as possible must be built in to facilitate rehabilitation and/or its adaptation for other uses (e.g. storage, workshops, depot, office). It is likely that the superstructure will be built over the dry well of a wet-well/dry-well pumping station.

When designing for ease of rehabilitation, the following points need to be considered carefully when assessing the need for future modification:

- compactness may subsequently create problems with space
- a building tailor-made for the plant leaves little scope for later upgrading
- concrete stairs, corbels, plinths, support beams, protective barriers etc. are strong, durable and maintenance-free, but steel is usually quite adequate and much easier to remodel or replace at a later date.

It is probable that tenders will be invited for the pumping plant from several pump manufacturers. It would be prudent, therefore, to ensure that the station can accommodate any of the systems offered by the tenderers, irrespective of which one is eventually chosen, and to go "one size up".

The use of steel, where practicable, in place of concrete will enable changes in layout to be carried out more easily than concrete, even though steel may be less attractive in appearance than concrete and require periodic painting. The use of galvanised steel should be considered.

5.11 LIFTING EQUIPMENT

The designer must decide upon the type and rating of lifting equipment to be provided at the pumping station. Clearly, provision must be made for reasonably major maintenance to be carried out, involving the dismantling of the pump and removal of the motor and gearbox. The manufacturer can advise on the correct rating of the lifting equipment. It would probably not be appropriate to provide lifting gear to enable the complete pump to be removed, as this would only be required during installation and rehabilitation. However, it is relevant to consider how the pumps will be lifted when they are installed or when they are removed during rehabilitation. It should be noted that permanently installed lifting equipment requires statutory annual testing. If a mobile crane is to be used, this may require the roof to be removed or suitable hatches to be provided, which will then have to be designed accordingly.

5.12 PUMPING MAINS

The design of the pumping mains is an important and integral aspect of the pumping system; see Sections 2.2 and 3.8. Consequently, initial design work for potential future rehabilitation of the pumping station should include a detailed review of the likely operating ranges of the pumping main(s), particularly if increased flows are anticipated.

During the actual design of the rehabilitation work in the future, the size and condition of the pumping main(s) should be reviewed.

5.13 CASE HISTORIES

Case Study 7 – Replacement of starters, electrodes, distribution and rewiring; improving access by enlarging access shafts; and providing ventilation via duct:

- problems: poor access and safety
- designer: in-house
- work by: own workforce.

Flow maintained by keeping one pump operating throughout.

Case Study 8 – Change in use of area required replacement of a single centrifugal pump with two submersibles:

- problems: lack of space
  poor access
- designer: in-house
- work by: own workforce, for the removal of the old pump and installation of lifting beam; installation of new plant by M&E contractor.

When the design of the pumping main in this example was reviewed, it was accepted that the small (100 mm) rising main would so restrict the output of the station that the pumps would have to operate on a duty/standby basis, as duty/assist would not be practicable. Current low flows enabled the sump inlet to be stopped off during pump changeover.

5.14 CHECKLIST

The following list, in the form of actions, sets out the key issues to be addressed in establishing designs for ease, speed and economy of rehabilitation.

Action	Reason	Section ref.	Checked	
Decide whether it is worth providing a structure with a longer design life than the plant	Whole-life cost	5.6	/	/
Consider how the structure can be designed to facilitate later modification. Possibilities include: providing additional space providing steel stairs and gangways providing weak concrete with rendering in benchings	Whole-life cost	5.6	/	/
Check whether there is room on the site for expansion of the structure or subsequent construction of a duplicate pumping station	Layout	5.9	/	/
Ascertain the kind of lifting equipment required and whether it is feasible to provide this as a permanent fixture at the pumping station	Whole-life cost	5.11	/	/
Consider whether there may be a future change in the pumping main(s) regime	Plant selection	5.3 5.12	/	/

6 Balancing whole-life costs

6.1 INTRODUCTION

There are several sources of finance involved in the construction and operation of pumping stations. If a water service company is building the station, construction costs will come from its capital budget, whereas the operating costs will be covered from its revenue budget. At any given time there may be constraints on these budgets. For example, it may be considered expedient to reduce the capital costs to ensure that the pumping station is built, even though this may increase operating costs. Alternatively, the priority may be to keep annual operating and/or maintenance costs to a minimum.

Quite often, a pumping station is built by a developer to serve a new development and subsequently adopted by the water service company. In this situation the construction costs are normally met by the developer and the cost of operating and maintaining the station is covered by the water service company. For this reason, the water service companies will not adopt a pumping station unless it complies with certain requirements.

Although the minimum requirements for the pumping station are those set out in *Sewers for adoption*[11], almost every water service company has its own, more stringent, requirements that developers must satisfy. The primary intention of these requirements, which may cover matters such as protective coatings, telemetry etc., is to keep the operating and maintenance costs to a minimum. The water service companies are responsible for the operation and maintenance costs for the life of the plant, together with potential difficulties when plant renewal or pumping station rehabilitation is required. As a result the special (additional) requirements of the water service companies vary widely due to variations in long-term views. In one example, the cost of a control panel to the water company's specification was 2.4 times the cost of a panel that complied with the requirements of *Sewers for adoption* 3rd edition[13].

The 4th edition of *Sewers for adoption*[11] (published during the production of this book) should help to overcome these problems. It covers pumping stations in far more detail than its predecessor, and sets out requirements for pumps, motors, control equipment and telemetry. It is hoped that these will be accepted, by all parties, as the standard requirements for the majority of new sewage pumping stations. This will enable developers to produce standard pumping station designs, providing a consistent quality with associated cost advantages, both to the developer with respect to capital costs and to the water service companies with respect to operating and maintenance costs.

6.2 WHOLE-LIFE COSTS

When carrying out an assessment of different alternatives for a new pumping station, the whole-life costs for different options ideally should be evaluated and compared. This is usually done by means of a net present value analysis.

For example, a new pumping station for a small development might be either a submersible pumping station with a 100 mm pumping main to the nearest point on the sewerage system, or an above-sewage pumping station with a progressive cavity pump and a macerator, and a small-bore polyethylene pumping main. The latter is likely to have reduced capital costs, particularly in respect to the pumping main, which will be relatively cheap to install. However, the higher velocities in the main and the resulting higher pumping head may increase operating costs, even allowing for the greater efficiency of positive displacement pumps compared with centrifugals. Thus a net present value analysis will enable the most cost-effective solution to be determined. A further option is to consider the use of submersible pumps with proprietary cutters or grinders.

There is a tendency among the various water utility companies to use different asset lives for M&E plant, e.g. 10, 15 or 20 years. In some cases, the life of the plant may be based on the total number of operating hours rather than a fixed period. These variations can have a significant effect on the calculated net present values and their relative importance in the estimation of whole-life costs. Replacement of individual components, such as bearings and seals should be allowed for in the estimation of maintenance costs. The adoption of standard asset lives would be of benefit to plant suppliers and developers in that they will be able to adopt standard criteria for the selection of the type of pumping station and the plant to be installed in it.

6.3 CAPITAL COSTS

Capital costs for pumping station construction are relatively easy to obtain. Data on civil costs is readily available in various forms, and manufacturers are usually willing to provide quotations when requested. While care should always be exercised, particularly with "budget quotations", a reasonably accurate estimate of the construction costs can be prepared.

Based on returns to questionnaires, it was apparent that where pumping stations were designed in-house by a water service company (or its agent), the plant costs are approximately one-third of the total construction costs, and the design costs are about 13% of the construction costs. Where the pumping station was designed and constructed by a developer, the M&E costs were significantly higher, around 50% of the construction costs, probably reflecting economies in materials and labour when the station formed part of larger works. In this case, the design costs were around 7 to 8% of the construction costs.

6.4 OPERATING COSTS

6.4.1 Power

Normally, designers estimate operating costs by calculating the work done by the pumps in a year, converting it into units of electricity used, and then multiplying the answer by the appropriate charge from the local electricity board's tariff. This, however, ignores other annual costs such as rates, telephone, water and power for other purposes such as heating and lighting.

Returns to the questionnaires indicated that operating costs were probably around 50% to 100% higher than the basic electricity tariff.

Selection of an electricity tariff can have a significant effect on the operating costs of a pumping station. Tariffs can be complicated issues and need to be appropriate to the size, nature and usage of the load. In general the nature of sewage pumping allows little scope for the use of off-peak tariffs.

Consultation with the regional electricity company (REC) is of paramount importance to ensure that the correct tariff is applied. Industrial tariff agreements may apply for a fixed period of up to five years. It is often not possible to modify a tariff agreement during this period; if it is, there is usually a financial penalty.

Each REC has its own range of tariffs, although the structure of tariffs is generally the same. The following is an outline of the typical criteria that are applied when selecting an industrial tariff:

- specify the type of supply required, i.e. high-voltage (11 kV) or low-voltage (240/415 V). For the majority of installations considered within this report the supply will be at low voltage
- decide whether the supply is to be billed quarterly or monthly. Quarterly-billed tariffs will typically only apply to loads of less than 50 kVA
- consider whether a general-purpose tariff or a maximum demand tariff is more appropriate. The general-purpose tariff is typically applicable to loads of less than 50 kVA or an annual consumption of less than 60,000 units. The maximum demand tariff is designed to ensure that the load demand does not exceed a declared maximum (in any half-hour period). The supply availability capacity is matched, in bands, to the maximum expected load demand; any additional consumption above the band declared maximum is penalised. For most installations covered by this report a maximum demand tariff is likely to be the most suitable
- determine whether a day/night tariff would be beneficial. Unit charges are cheaper during some hours (typically night-time) than others. This type of tariff will be beneficial only where pumping periods can be controlled (e.g. pumped storage schemes). For installations covered within this report the applicability of a day/night tariff should be discussed with the REC.

As tariff structures can appear complicated at first sight the following outlines the typical make-up of a maximum demand tariff:

- monthly charge – a fixed monthly charge for the service
- supply availability charge – a monthly charge for each kVA of maximum capacity held available by the REC
- maximum demand charge – a charge for each kVA of demand made in specific months, when the declared maximum demand is exceeded. Typically a charge is levied for consumption in the winter months (in an attempt to discourage consumption during these periods of higher demand)
- unit charge – a flat-rate charge for each unit consumed
- fuel price adjustment – an adjustment charge to take account of fluctuating fuel costs to the generating/regional electricity companies.

Clearly, it is essential to discuss tariff selection with the REC as soon as the load characteristics are known. The REC will provide an assessment of the running cost for a particular installation against a number of available tariffs and advise on the selection of the most suitable.

As electrical energy may be purchased from any REC, the differences in the various tariffs should be taken into consideration. However, for the typical pumping stations considered within this report (unless the station is located on the border between two or more REC areas) the additional cost of transporting the energy to a particular site is likely to negate any benefits of 'shopping around'.

The design of the pumping main and the pumping station pipework has implications for overall power consumption – see Sections 3.8 and 3.9.

6.4.2 Maintenance

This is another area where costs are difficult to determine accurately and objectively. Manufacturers may be asked for information regarding maintenance costs, but this often proves to be unreliable, as it only covers mechanical and electrical maintenance and does not take into account routine cleaning of the sump etc. or the variety of practices between operators.

Generally, annual maintenance costs should allow for the following:

- routine maintenance – covering all the items of work normally carried out by the maintenance gangs during their regular visits to the station, which range from bi-weekly to monthly
- planned maintenance (M&E) – covering routine servicing and inspection of the pumps, motors and control gear
- reactive maintenance (M&E) – covering emergency repairs and work resulting from the detection of a fault (or potential fault) during planned maintenance.

While routine maintenance is normally carried out by the pumping station operator's staff or a contractor employed for this type of work, planned and reactive maintenance would normally be carried out by a specialist M&E contractor, either on an *ad-hoc* basis or under some form of annual maintenance contract. The basic annual contract would normally cover planned maintenance, with reactive maintenance being done on a basis of need, at extra cost. A "comprehensive" annual contract would cover both planned and reactive maintenance, and the annual cost would be based on the manufacturer's estimate of the reactive maintenance that may reasonably be necessary during the projected life of the plant.

Based on a survey of such contracts, one manufacturer suggests that the annual maintenance cost over a 20-year plant life would be approximately equal to 5% of the cost of supplying and installing the original plant. It is probable that this figure should be at least doubled in order to include the cost of routine maintenance.

6.5 RECORDS

It is not clear what records operators keep and how they relate to individual pumping stations; indeed, practices vary widely in this respect. Since routine maintenance is almost invariably carried out by travelling maintenance gangs, it is important that they keep a record of the time spent at each station. They should also keep a record of the time spent on maintaining the plant as distinct from the time spent on cleaning down the sump, checking the operation of the valves etc. Travelling time should be allocated fairly, or kept separate. It is possible for a pumping station to have apparently very high maintenance costs simply because it is located a long distance from base.

There may be some benefit, in respect to budget control, in taking out a plant maintenance contract with a plant maintenance contractor or a plant manufacturer. In some cases, for an agreed annual sum, they will accept the risk and cover for all reactive maintenance. The operator could then use a direct labour gang or a contractor to carry out the routine station cleaning and maintenance, which means that their operatives would not be required to have any knowledge of mechanical or electrical matters. Not only would this facilitate the keeping of more accurate records, it could result in a more cost-effective maintenance operation.

It is not clear how operators allocate ancillary operating costs, and whether they are aggregated together or allocated to individual pumping stations. These ancillaries include water supplies, telephone/telemetry links and rates. Electricity power supplies will be metered at the station and will include ancillary power uses such as heating and lighting.

It is also not clear whether, and how, management costs are allocated to individual pumping stations.

Although every pumping station will have its own individual requirements, it will be impossible to obtain a precise estimate of the operation costs of any new pumping station. The maintenance of accurate records will enable a picture to emerge, so that a reasonable estimate can be made of operating and maintenance costs for pumping stations of different types and sizes.

6.6 OPERATIONAL PRACTICES

One important reason why operating and maintenance costs (particularly the latter) vary widely between is that the operators all follow different practices. For example, the frequency of maintenance visits ranges from daily, through bi-weekly, weekly, fortnightly and monthly, to every six weeks. Sometimes this is tailored to the needs of the pumping station, whereas in other cases it is based on a management decision. Advanced telemetry may make it possible to reduce the frequency of maintenance visits as it will enable any problems occurring between visits to be detected through the telemetry system. These issues will have varying effects on the operating and maintenance costs of each station.

It is therefore important that each operator builds up a record, based on his own operational practices, which can be used by the designer in his assessment of the options and to take into account in his design.

6.7 CONCLUSIONS

Whereas capital costs and power costs for pumping are reasonably simple to estimate, there are other operating and maintenance costs that will have a significant bearing on the actual whole-life costs of the station, which are difficult to determine at the design stage.

Some of these costs relate to the operator's normal practices, while others will relate to the type of pumping station or the nature of the sewage to be pumped.

It is in the interests of the operators to keep accurate and detailed records of operating and maintenance costs for each pumping station, and to liaise with the designers of new pumping stations to ensure that correct assessments of these costs have been made in respect of any decisions made at the design stage.

6.8 CHECKLIST

The following list, in the form of actions, sets out the steps to be followed in establishing the whole-life costs of the plant and selecting the appropriate type of pumping station to be provided.

Action	Reason	Section ref.	Checked	
General				
Determine the design life of the plant	Plant selection	6.2	/	/
Determine the design life of the civil structure	Type of installation	6.3	/	/
Estimate the construction cost	Cost comparisons	6.3	/	/
Obtain data on operation and maintenance costs	Operating practices	6.4 6.6	/	/
Calculate the likely cost of rehabilitation (if applicable)	Maintenance	6.4.2	/	/
Decide what discount rate(s) should be adopted for calculating the net present values and carrying out any sensitivity analysis	Financial evaluation	6.2	/	/
Estimate the whole-life costs or net present values for the different design options	Type of installation	6.2	/	/
Compare the above costs for the different design options	Type of installation	6.2	/	/
Take into account the constraints on capital expenditure and O&M costs	Operating costs	6.4	/	/
Establish which economic design option complies with the above constraints	Type of installation	6.7	/	/

7 Summary and checklists

7.1 INTRODUCTION

This book is concerned with design, and although this has been considered under the different headings of construction and plant installation, routine operation and maintenance, major maintenance and repair and rehabilitation, the key features of a good design for buildability and maintenance are common to all four topics. These are:

- simplicity of design
- adequate health and safety provisions
- minimum maintenance (including landscaping)
- good access
- provision of adequate lifting equipment
- standardisation of plant and equipment.

The key to achieving all these things is good communication between all the parties involved in the design, construction and operation of the station, including the plant and equipment suppliers and installers.

7.2 RECOMMENDATIONS FOR GOOD PRACTICE

The major concerns for a pumping station designer are achieving reliability and value for money. In the past the approach has been to complete the design in isolation and then obtain competitive tenders from plant and civil contractors.

Communication with other members of the team (designer, supplier, contractor, operator) can provide benefits, as explained elsewhere in this volume. The cost benefits include:

- use of standard items of plant and equipment, with associated savings in cost and improved availability of spares
- standardisation throughout the operating district, with improved training of maintenance gangs, more efficient maintenance and emergency responses, need for a smaller range of spares and greater scope for keeping a stock of the more frequently used spares at base or in the van, and greater potential for exchange of units of plant in an emergency
- proper provisions for access, based on plant needs
- sensible and adequate provisions for lifting equipment
- sensible and adequate provisions for health and safety during construction and maintenance operations
- an appropriate balance between capital cost and operating costs.

The key recommendation for good practice is to involve the whole team as soon as possible, with full, detailed and frequent dialogue between all the team members. Discussions should be properly recorded in correspondence or minutes of meetings, with any minutes agreed by all parties. To some extent, such discussions are a natural outcome of the CDM Regulations[1], so it will not require a great deal of effort to extend them to encompass all aspects of the design. The loss of the competitive edge from the tendering process will be fully compensated for by the other savings outlined above, coupled with fewer disputes, better allocation of responsibilities and shorter, more concise specifications. "Cheapness" will largely be eliminated in favour of greater value for money coupled with significant improvements in ease of construction and all aspects of operation and maintenance. This approach will be further assisted by the increased use of framework agreements and partnering arrangements. The following checklist should assist in team discussions.

7.3 CONSOLIDATED CHECKLIST

The following checklist presents the contents of the checklists at the ends of each of the preceding sections in a consolidated form:

Action	Reason	Section ref.	Checked	
General				
Establish the proposed design horizon for the plant	Whole-life cost	2.2	/	/
Determine the proposed design horizon for the civil structure	Whole-life cost	2.2	/	/
Consider site constraints on the type of pumping station, e.g. access, topography, aesthetics, etc.	Layout	2.4	/	/
Consider whether a package-type pumping station would be appropriate	Small installations	2.5.5	/	/
Evaluate the aesthetic requirements of the locality	Statutory requirement	2.2 2.4	/	/
Decide on the need for a superstructure	Type of installation	2.5.3 2.5.4	/	/
Function				
Consider the type of sewerage system, i.e. whether it is separate or combined	Characteristic of flow	2.2	/	/
Determine the present dry weather flow	Design basis	2.2	/	/
Estimate the dry weather flow at the design horizon for the plant	Whole-life cost	2.2	/	/
Calculate the probable changes in flow conditions between the design horizon of the plant and the design horizon for the civil structure	Whole-life cost	2.2	/	/
Quantify the maximum flow to be pumped	Plant selection	2.5	/	/
Estimate the probable maximum flow to the pumping station	Plant selection	2.5	/	/
Determine the elevation of the discharge point compared with the pumping station	Determine static head	2.2	/	/
Measure the flow velocity in the pumping main	Determine friction head	2.2	/	/
Calculate the size and length of the pumping main and its discharge point	Determine friction head	2.2	/	/
Determine the static lift required and the total head on the pumps	Plant selection	2.5	/	/
Assess which types of pumps are suitable for the required duty	Plant selection	2.5	/	/
Decide whether a storm or emergency overflow is necessary and where it will discharge	Statutory requirement	2.2	/	/

Action	Reason	Section ref.	Checked	
Assess what power supply is available and its characteristics in respect of permissible starting currents	Services	2.8	/	/
Consider the best way of safeguarding the power supply, as well as the noise and exhaust gas emission implications of any standby power generation required	Services	2.8	/	/
Decide on the cable routing	Services	2.8	/	/
Determine the availability of water supplies	Services	2.8	/	/
Investigate the requirements for heating, lighting and ventilation, socket outlets, intruder alarms, lightning protection	Services	2.8	/	/
Determine the need for, and method of, grit removal and disposal at the pumping station. If possible, grit should be retained in the sewage flow and separated out at the sewage treatment plant	Construction / plant selection	2.4 2.5	/	/
Establish the requirements for odour-control measures at the pumping station (particularly if grit removal is carried out)	Environmental requirements	2.2	/	/
Evaluate the telemetry requirements	Communication	2.9	/	/
Buildability				
Decide on suitable locations for the site in relation to the gravity sewer and the pumping main	Layout	2.2	/	/
When choosing the location, consider access for construction plant and equipment, and the delivery and storage of plant, equipment and materials	Layout	2.2	/	/
Ensure that a suitable power supply is available	Services	2.8	/	/
Consider the ground conditions at the site, height of the water table and risk of flooding at the site	Construction technique	2.4	/	/
Consider the sewage flow implications during construction	Construction safety	2.4	/	/
Examine the likely advantages of using standard precast or prefabricated items, e.g. precast concrete segments for sump construction and prefabricated GRP buildings for the superstructure or control kiosk	Whole-life cost	2.2	/	/
Assess the likely benefits of adjusting the design in order to accommodate manufacturers' standard items of plant and equipment	Whole-life cost	2.2	/	/

Action	Reason	Section ref.	Checked	
Statutory requirements				
Ascertain whether a consent is needed for an overflow	Statutory requirement	2.2 2.7	/	/
Establish whether an overflow would need to be screened to minimise discharge of solids	Statutory requirement	2.2 2.7	/	/
Apply for planning permission	Statutory requirement	2.2 2.7	/	/
Examine the conditions applied by the planning authority to the development	Statutory requirement	2.2 2.7	/	/
Appoint a planning supervisor as required under the CDM Regulations[1]	CDM	2.10	/	/
If the pumping station is to be adopted on completion, take into account the required design standards – e.g. *Sewers for adoption*, 4th edition, or water service company design specifications	Statutory requirement	2.2 2.10	/	/
Communication				
If the pumping station is to be adopted on completion, ensure the water service company is consulted at an early stage	CDM	2.9 2.10	/	/

Initiate discussions with the following:
    the plant operator
    the pump manufacturer
    the M&E installation contractors
    the civil contractor
	Maintenance Construction Plant selection	2.9	/	/

Ensure that the CDM communication requirements have been satisfied	Statutory requirement	2.2	/	/
Access to the site				
Decide on suitable locations in relation to access for construction plant and equipment, and the delivery and storage of plant, equipment and materials	Layout	3.2	/	/
Ensure permanent vehicular access to the site	Layout	3.2	/	/
Provide adequate parking and turning space	Layout	3.2	/	/
Allow room for a vehicle-mounted washwater booster, a mobile pump or a generator to stand close to the pumping station	Sump cleansing Emergency power	3.10 3.11	/	/
Plant design				
Ensure that the plant is suitable for the required duty and capable of passing solids of an appropriate size	Reliability	3.4	/	/

Action	Reason	Section ref.	Checked	
Check that the pipework is designed to minimise the risk of blockages and to ease clearing of any blockages that might occur	Operational problems	3.5.1	/	/
Check that the plant is suitable for any relevant hazardous area classifications	Safety	3.14	/	/
Consult the pump manufacturer about measures for achieving optimum reliability by reducing the frequency for major maintenance	Reliability	4.4 4.5	/	/
Consult the pump manufacturer about lifting requirements for maintenance of the plant	Plant selection	4.2	/	/
Consult the pump manufacturer about access, working space and ventilation requirements	Layout	4.2 4.4	/	/
If the pumping station is to be adopted on completion, consult the water service company at an early stage	CDM	2.9 2.10	/	/
Rationalisation				
Investigate the scope for standardisation to improve spares availability, interchangeability of parts or whole units, speed and efficiency of maintenance work	Whole-life cost	4.8	/	/
Decide on the minimum level of complexity of plant and control equipment necessary for the reliable operation of the plant and consider whether there is any justification for exceeding this	Plant selection	3.7	/	/
Check that the complexity of the plant to be installed is appropriate for the anticipated skills of the operation and maintenance team	Operator training	4.8	/	/
Consider whether maintenance operations can be made more efficient by promoting uniformity/ standardisation of plant and equipment within the operating district	Plant selection	3.7	/	/
Civils design				
Check the sump capacity is suited to the application and pump starting limitations	Hydraulic problems	3.6	/	/
Ensure the sump design is the best possible to avoid hydraulic problems with the pumps	Reliability	3.6	/	/
Minimise air entrainment	Prevent cavitation	3.6	/	/
Evaluate the sump design with regard to its efficiency in minimising the deposition and build-up of grit and other solids	Reliability	3.6	/	/

Action	Reason	Section ref.	Checked	
Consider whether there may be a future change in the pumping main(s)/regime	Plant selection	5.3 5.12	/	/
Consider how the structure can be designed to facilitate later modification. Possibilities include: providing additional space providing steel stairs and gangways providing weak concrete with rendering in benchings	Whole-life cost	5.11	/	/
Access, working space and ventilation				
Ensure that access openings are large enough to facilitate the installation and removal of plant and equipment	Maintenance	3.10	/	/
Provide adequate access to the sump for cleaning, with appropriate numbers and sizes of covers, access ladders/stairways	Maintenance safety	3.2 3.10	/	/
Decide whether pumps are to be removed for inspection and repair or whether this work will be carried out in-situ	Consider hazards	4.3	/	/
Provide adequate access to all plant and equipment	Safety	3.10	/	/
Make sure that there is sufficient space for the work to be carried out	Layout	4.3	/	/
Consider what additional ventilation will be required while people are working at the station and the provisions that should be made for this in the design	Safety	4.3	/	/
Ensure that the pipework arrangement facilitates cleaning/unblocking of delivery pipework, if required	Layout	3.2 3.9	/	/
Facilities to be provided				
Determine the kind of lifting equipment that is required and its suitability for the pumping station. If portable lifting plant or equipment is to be used, provide adequate access and consider whether it can be provided as a permanent fixture at the pumping station	Maintenance	3.10	/	/
Investigate the possibility of employing a pressure washwater system for efficient cleaning of the sump. Consider whether a booster unit should be installed at the station or carried on the maintenance vehicle	Maintenance	3.2 3.10	/	/

Action	Reason	Section ref.	Checked	
Decide whether the major maintenance is to be carried out by in-house engineers, a general maintenance contractor (with specialists employed when required), a specialist M&E contractor, or the pump manufacturer	Operating costs	4.7	/	/
Specify what equipment is to be carried by the maintenance engineers and what should be provided at the station. This relates particularly to lifting equipment	Maintenance operations	4.7	/	/
Establish what loads the station lifting equipment will be required to lift	Safety	4.7	/	/

Determine how the pumps are to be isolated and flows maintained during maintenance. Options include:
    isolating valves on suction and delivery pipework
    divided sump with emergency facilities
    provision of facilities for over pumping
    standby power supplies
	Reliability/maintaining operations	4.5	/	/

Provide adequate access to the sump for cleaning, with appropriate numbers and sizes of covers, access ladders/stairways	Maintenance safety	3.2 3.10	/	/
Ensure there is adequate ventilation to enable cleaning operations to be carried out easily and safely	Maintenance safety	3.2 3.10	/	/
Consider fitting automatic flushing valves to pump volutes	Plant selection	3.5.1 3.7	/	/
Ensure that the pipework arrangement facilitates cleaning/unblocking of delivery pipework, if required	Layout	3.2 3.9	/	/
Health and safety				
Comply with the CDM Regulations[1] and prepare a health and safety file	Statutory requirement	3.4 3.15	/	/
Carry out a hazardous areas analysis and ensure that plant is suitably rated	Safety/reliability	3.4 3.14	/	/
Carry out a hazards analysis and hazardous operations study (HAZANS and HAZOPS) and take account of the recommendations in the design	Safety/reliability	3.4 3.14	/	/
Provide adequate warning notices at the station regarding hazards and confined spaces	Safety	3.14	/	/

Action	Reason	Section ref.	Checked	
Establish what safety provisions are needed, e.g. emergency cut-outs, guard rails etc.	Control/safety	4.6 4.8	/	/
Decide whether it is appropriate to provide sanitary, washing and/or messing facilities	Amenities	3.2	/	/
Communication facilities (ICA etc)				
Determine the appropriate level of telemetry for the pumping station	Telemetry	3.13	/	/
Establish what emergency alarm facilities are required	Telemetry	3.13	/	/
Decide whether signals are to be communicated by telephone, dedicated landline or radio. If by radio, establish whether there are any constraints on the type, size and height of the aerial	Telemetry	3.13	/	/
Vandalism				
Evaluate the level of risk of vandalism at this site	Security	3.5.2	/	/
Determine appropriate measures to minimise damage in respect of site security, lighting, security cameras, types and design of superstructure (avoidance of windows) and construction materials (resistance to damage and/or ease of cleaning or overpainting)	Security	3.5.2	/	/
Manuals				
Provide a comprehensive operation and maintenance manual setting out full procedures for the station's routine maintenance including health and safety requirements	Communication	3.15	/	/
Ensure that the manual provides sufficient information in the right form to enable the work to be carried out	Communication	3.15	/	/
Check that all special provisions and procedures for health and safety during major maintenance are set out clearly in the O&M manual. Make sure that these provisions are identified in the health and safety file	Safety	4.8	/	/
Check that the manual highlights the need for regular certification of lifting devices	Safety	3.15	/	/
Decide where to keep the manual	Communication	3.15	/	/
Ensure that the station maintenance staff have ready access to the O&M manual	Communication/Safety	3.15	/	/
Costs				
Determine the plant's design life	Plant selection	6.2	/	/

Action	Reason	Section ref.	Checked	
Determine the design life of the civil structure	Type of installation	6.3	/	/
Estimate the construction cost	Cost comparisons	6.3	/	/
Obtain data on operation and maintenance costs	Operating practices		/	/
Calculate the likely cost of rehabilitation (if applicable)	Maintenance	6.4.2	/	/
Decide what discount rate(s) should be adopted for calculating the net present values and carrying out any sensitivity analysis	Financial evaluation	6.2	/	/
Estimate the whole-life costs or net present values for the different design options	Type of installation	6.2	/	/
Compare the above costs for the different design options	Type of installation	6.2	/	/
Take into account the constraints on capital expenditure and O&M costs	Operating costs	6.4	/	/
Establish which economic design option complies with the above constraints	Type of installation	6.7	/	/

8 Recommendations for future work

Generally, data on capital costs is readily available, either through one of the standard databases or as historical data compiled by the firm that is carrying out the design. However, in carrying out whole-life costings, there are difficulties in the area of operation and maintenance costs, where very little information is readily available in a usable form. Where such information does exist, it tends to be extremely variable. There are two main reasons for this:

1. Practices vary considerably between operators, e.g. the frequency of routine maintenance visits and the scope of the work carried out during such a visit.
2. The extent and detail of records regarding the time and costs of maintenance vary between operators.

It is recommended that a standardised recording scheme is devised, and for the costs incurred by various operators, and/or the time involved in maintenance work, to be recorded. This would have two useful results:

1. It would identify those operators that may be spending more time and effort on maintenance than the average, and indicate areas where savings could be made.
2. It would provide a database of reliable historical costs, which could be used beneficially for project appraisals.

Almost all the water service companies have requirements that differ from, and often far exceed, those of *Sewers for adoption*. This prevents designers from producing standard designs and results in increased costs. Water plcs should seek to achieve common "additional requirements" to those included in *Sewers for adoption*, as this will enable standardisation of designs to be achieved, leading to overall economy in both capital and operating costs.

Appendix A1 Consultations

Consultations were held with 11 contacts, of which eight were water companies or agencies acting on their behalf, and three were pump manufacturers.

Questionnaires were sent out, with a view to the contacts completing it before being visited by the research contractor. In the event, few of the contacts had completed the questionnaire in time for the meeting. Only eight questionnaires were completed, out of 16 sent out.

Key points arising from the consultations and questionnaire responses are as follows:

- communication between designer, contractor/supplier/installer and operator is vital
- good housekeeping is essential throughout the life of the pumping station. This demands a level of funding that is appropriate to the individual pumping station
- good sump design has the greatest impact on pumping station operation
- access and lifting arrangements have the greatest impact on maintenance activities
- historic costing is rarely available in an accessible, reusable form
- the appropriate level of control and monitoring facilities requires strategic re-examination
- rationalisation of pumping station design, particularly in respect of plant and equipment (especially control panels) would greatly assist efficient and cost-effective maintenance, and improve response in emergencies
- historic records of simple parameters, such as motor current, provide a good indication of the present status of the pumping equipment. Any unexplained changes from the norm should be investigated.

Appendix A2 Cost data derived from consultations

A2.1 CAPITAL COSTS

Since most of the respondents to the questionnaires were pumping station operators, very little capital cost information was available. However, two water utility companies and one developer gave information on a total of six pumping stations. A digest of this data is given in Table A2.1.

For the utility companies (or their agents) the ratio of [design cost : civils cost : plant cost] is reasonably constant and averages out at [1 : 5.28 : 2.34]. Thus the plant cost is approximately 31% (say one-third) of the construction cost. For the pumping stations designed and constructed by a developer, the ratio is quite different, the plant cost varying between 45% and 66% of the construction cost depending on whether or not there is a superstructure. Design costs for the utility companies varied between 11% and 17.6%, averaging about 13%. Design costs for the developer were lower at 8.6% and 6.4%, averaging out at just over 7%.

An attempt was made to relate the costs to installed power, but the scatter was so wide that no relationship could be derived from the available samples. Relating costs to the maximum output in litres per second was a little more successful, particularly for Pumping Stations A to D (those designed in-house by the utility companies), such that when unit costs in £ per l/s output were plotted against $\log_{10}$(max output in l/s) something approximating to a family of straight lines was obtained. This was applicable only up to about 25 l/s output, as the unit cost approached zero at about 32 l/s. However, the costs for Pumping Stations E and F did not fall on these lines.

A2.2 OPERATING COSTS

Normally, designers estimate operating costs by calculating the work done by the pumps in a year, converting it into units of electricity used, and then multiplying the answer by the appropriate charge from the local electricity board's tariff. This, though, ignores other annual costs such rates, telephone, water, and power for other purposes such as heating and lighting.

A few respondents to the questionnaires completed the Schedule of O&M Costs. An analysis of these shows a wide variety of operating costs – see Table A2.2. One respondent, Water Company No 3, gave annual power costs together with the numbers of units used. In this case the cost per unit generally varied from 6p to 17p, with 37p quoted in one case. The average power cost in this district was 12p per unit.

It is difficult to know how the calculated cost per kWh in the table compares with the actual tariffs quoted by the electricity undertakers, but it is probable that something needs to be added to cover extras over the basic cost of pumping. If the tariff normally used to estimate operating costs is 6 to 7p, then it is probable that it should be increased by about 50% to give operating costs of the order of those given in Table 4. Tariffs, and methods of calculating them, vary widely across the electricity utility companies, and these will also vary with conditions of use. It is therefore very difficult to estimate reliably the operating costs of a pumping station, but it is suggested that a more accurate figure will be obtained if the quoted tariff is enhanced by between 50 and 100%.

A2.3 MAINTENANCE COSTS

This is another area where costs are difficult to determine accurately and objectively. Manufacturers may be asked for information regarding maintenance costs, but this often proves to be unreliable, as it only covers mechanical and electrical maintenance and does not take into account routine cleaning of the sump etc. nor the variety of practices between operators.

In Table A2.2, the annual maintenance costs for Water Company No 1 vary from £300 pa to £800 pa, with the latter figure applying to about 50% of the stations quoted. For Water Company No 2, the maintenance costs vary from about £240 to just under £2000 pa. It is not clear exactly whether the above figures include, for example, regular servicing of the pumps, motors and control gear (which usually takes place about twice a year), or just the regular routine maintenance operations of sump cleaning, checking valve movement etc.

Water Company No 3 breaks its maintenance costs down into three categories:

1. Routine maintenance – covering all the items of work normally carried out by the maintenance gangs during their regular visits to the station, which range from bi-weekly to monthly. The costs varied quite widely, but averaged out at £1235 pa.
2. Planned maintenance (M&E) – covering routine servicing of the pumps, motors and control gear. These costs varied by a factor of 7, but averaged out at about £226 pa per station.
3. Reactive maintenance (M&E) – covering emergency repairs and sometimes quite major work resulting from the detection of a fault during maintenance. These figures varied widely from £12.25 to just under £15 000 during the relevant year. It is hoped that some of these costs would not be repeatable. The average cost per station was about £1058. It is noted that every pumping station in the district required some degree of reactive maintenance during the year in question.

It would not be unreasonable to consider that the major items of reactive maintenance comprise "major maintenance" as discussed in Chapter 4, as one would hope that they do not occur often. The status of the minor items of reactive maintenance is unclear: they may cover occasional parts replacement, carried out during a planned maintenance visit, but not included in the planned maintenance budget. However, if the average annual plant maintenance costs are taken as the sum of the planned maintenance and the reactive maintenance, the average annual plant maintenance cost comes to £1284.

In estimating maintenance costs for a pumping station, it is necessary to take into account routine maintenance, planned plant maintenance and a reasonable prediction of probable reactive maintenance. This, of course, is the basis of maintenance contracts provided by pump manufacturers/maintenance contractors. The basic contract would cover planned maintenance, i.e. regular lubrication and inspection of the plant, which would take place once or twice a year.

A manufacturer of macerator/progressive cavity pump units has given some indicative maintenance costs over a 10-year period, which average out at £1410 per annum for a pumping station with an output of about 10 l/s. It is assumed that these costs would be similar to those charged for a comprehensive maintenance contract and thus would be equal to the sum of the planned maintenance and anticipated reactive maintenance.

The same manufacturer suggests that, based on a survey carried out of some 250 sewage pump maintenance contracts, the average annual cost for ongoing maintenance during an assumed life of 20 years would be about 5% of the cost of providing and installing the M&E plant at the pumping station. Again, it is assumed that this would cover both planned maintenance and anticipated reactive maintenance.

Table A2.1 *Design and construction costs for six submersible pumping stations, based on completed questionnaires*

Pumping station	A	B	C	D	E	F
Max output (l/s)	25.0	11.0	13.0	10.0	6.0	8.7
Number of pumps	2	2	2	2	2	2
Motor size (kW)	7.5	1.6	16.0	2.7	1.3	2.5
Size of rising main (mm)	100	40*	100	100	100	100
Costs, expressed as a ratio with design costs taken as unity						
Design cost	1	1	1	1	1	1
Civils cost	5	6.5	4.6	5.03	4	8.67**
Plant cost	2.5	2.5	1.07	3.3	7.67	7
Plant cost per installed kW	£667	£1563	£972	£3889	£8846	£4200
Total cost per installed kW	£2200	£6250	£2800	£10 926	£14 615	£10 000
Total cost per l/s pumped	£1360	£1800	£6893	£5900	£6333	£5747

*   Pumps fitted with macerators
**   Pumping station provided with superstructure

Note: Pumping stations A to D were designed in-house by water utility companies or their agents; pumping stations E and F were designed and constructed by developers.

Table A2.2 *Operating and maintenance costs*

Water company	Name of pumping station	Rated capacity (l/s)	Total no. of pumps	Total installed power (kW)	Hours run per annum	Operating cost per annum (£)	Maintenance cost per annum (£)	Operating cost per kWh
Water Company No 1	PS1	144	2	37	1500	1400	800	£0.05
	PS2	162	2	37	1550	2000	800	£0.07
	PS3	26	2	6	4112	1500	800	£0.12
	PS4	50	2	11	3510	1300	800	£0.07
	PS5	21	2	14.8	3251	1650	300	£0.07
	PS6	17.4	2	11.2	1150	1000	300	£0.16
	PS7	16	2	42.6	1533	2150	300	£0.07
	PS8	27	2	2.2	5400	200	450	£0.03
	Averages	58	2	20	2751	1400	594	£0.05
Water Company No 2	PS1	6.8	2	25	16 698	4695	1426	£0.02
	PS2	25	2	15	252	1588	639	£0.84
	PS3	1.1	2	3.2	669	1070	1994	£1.00
	PS4	6	2	3.8	658	1509	915	£1.21
	PS5	4	2	8	970	2423	271	£0.62
	PS6	4	2	8	398	2309	241	£1.45
	PS7	100	2	41	122	720	400	£0.29
	Averages	21	2	15	2824	2045	841	£0.10
Water Company No 3	**Averages**	–	–	–	–	1832	2519	£0.12

Note: The hours run per annum for Water Company No 2's PS1 are very high and are equivalent to both duty and standby pumps operating almost continuously.

The maintenance costs for Water Company No 3 are made up as follows:
    Routine maintenance: £1235
    Planned maintenance: £226
    Reactive maintenance: £1058

Appendix A3 Literature review

A3.1 CIRIA PUBLICATIONS

Building Design Report: Practical buildability. By Stewart Adams (1989)

Defines the concept of buildability, and sets out the general principles of designing for buildability, illustrated with a large number of examples covering both strategic design and detailed design.

Special Publication 26: Buildability: an assessment (1983)

Good buildability leads to major cost benefits for clients, designers and builders. The achievement of good buildability depends on both designers and builders being able to see the whole construction process through each other's eyes. This report is a forerunner to *Practical buildability* above.

Report 121: Design of low-lift pumping stations with particular application to wastewater. By M J Prosser. Produced in conjunction with the British Hydromechanics Research Group Ltd (1992)

Covers the basic principles of pumping station design, including factors affecting the design process, pumping system arrangements, pump choice, operational requirements, electricity supply, operation, environmental impact, economic considerations and hydraulic design, and describes their particular application to the design of low-lift sewage pumping stations.

Report 122: Life cycle costing – a radical approach. By D J O Ferry and R Flanagan (1991)

Describes the principles of life-cycle costing and concludes that its optimum use is as a management tool over the fixed asset life cycle, with particular application to the control of overall plant and machinery costs. Life-cycle costing techniques are reviewed and explained in a detailed step-by-step process, and recommendations are made on their use and implementation in the construction industry.

Technical Note 132: Screenings and grit in sewage – removal, treatment and disposal. Phase 3: Storm water overflows and pumping stations (1988)

Deals with problems arising upstream of sewage treatment works, particularly at storm overflows and sewage pumping stations. Considers current practice, probable future practice, energy costs, financial costs, and makes recommendations about the use of screening and grit-removal equipment.

CIRIA with BHRA Fluid Engineering: The hydraulic design of pump sumps and intakes. By M J Prosser. Produced jointly by CIRIA and the British Hydromechanics Research Association (1977)

Outlines the problems involved in the design of a pump sump and intake. Recommends good practice for a variety of pumping stations, with preferred designs for suitable situations. Advice is given on the desirability of model tests, with guidance on the conduct of such tests.

A3.2   FOUNDATION FOR WATER RESEARCH

WSA/FWR Sewers and Water Mains Committee: A guide to sewerage operational practices (1991)

Covers planning, working practices and term contract documentation. Section C7 deals specifically with sewage pumping installations, and covers management objectives, cost-effective management, recommended procedures and working practices.

Report No FR 0452: Reliability and impact of failure at sewage pumping stations. By N Orman (1994)

The causes and consequences of sewage pumping station failure, and ways of preventing them, are investigated with particular reference to public power supply failure and rising main failure. It is concluded that only 2% of the many component failures each year result in either sewer flooding or emergency discharges to receiving waters. There is a very low rate of discharge or flooding from public supply failure, despite the fact that only 4% of installations are provided with permanent standby generators. However, while rising-main failures account for only 1% of component failures, these led to 35% of the incidents of sewer flooding or discharge to receiving waters. The design of rising mains should include an assessment of surge pressures, and the design of PVC and PE rising mains should take account of fatigue under surge pressures. Surge pressures should also be investigated when pumps are being replaced.

A3.3   WATER RESEARCH CENTRE

Sewers for adoption – a design and construction guide for developers (3rd edition). Published by the Water Research Centre on behalf of the Water Authorities Association (1989)

A design guide for private developers, setting out requirements for sewerage systems that are to be adopted by the water plcs. Generally regarded as the standard practice document for sewerage design and construction in the UK. Part 2, Section 2 covers pumping stations, and includes sections on wet wells, provision for access, pumps and motors, electrical installation, control panels, rising mains and maintenance.

The 4th edition was published in 1995 with an expanded and more comprehensive section on pumping stations.

Technical Report TR 170: Optimisation of sewage pumping. By G P Evans (1981)

Describes a new technique for controlling sewage pumping stations, developed by WRc, in which a computer is used to forecast the flow to the pumping station and to start and stop appropriate pumps, maximising the use of the storage capacity of the sewerage system upstream of the station. By taking into account electricity board tariff structures, savings in operating costs of up to 25% may be realised.

Technical Report TR 194: The application of computers at sewage treatment works, water treatment works and pumping stations – an annotated bibliography covering the period 1971 to 1982. By I M Hamilton (1983)

A worldwide review of relevant literature, divided into three sections: sewage treatment works, water treatment works, and pumping stations.

Processes TR 250: Sewage pumping – an energy-saving manual. By J A Hobson (1987)

Careful planning can reduce expensive oversizing of pumps. Additional lower rates of pumping should be considered for saving energy. Care should be taken to select pumps whose points of best efficiency match the best estimate of duty. In some cases, control of pumps, using system storage for balancing, can provide further energy savings and increase the use of off-peak electricity. Design teams must follow up designs to obtain feedback. Appendix A takes a new look at the economic sizing of rising mains.

A3.4 BRITISH STANDARDS

BS 8005: Part 2: 1987 – Sewerage. Part 2: Guide to pumping stations and pumping mains

Provides guidance on the components, appliances and design of pumping stations and pumping mains.

A3.5 WATER POLLUTION CONTROL FEDERATION (USA)

Manual of Practice No FD-4: Design of wastewater and stormwater pumping stations (1981)

Provides the design engineer with practical and proven guidelines for the design of wastewater and stormwater pumping stations. Covers the kinds of pumping stations that may be used in various situations, the selection and design of mechanical equipment, piping system layout and hydraulics, electrical design, and structural and architectural considerations.

A3.6 NATIONAL JOINT HEALTH AND SAFETY COMMITTEE FOR THE WATER SERVICE

Health and Safety Guideline No 2: Safe working in sewers and at sewage works (1979)

A guide for developing working procedures for individual processes and/or tasks. Draws attention to the common and foreseeable hazards inherent in sewage work, the various types of equipment available, the recommended precautions to be taken, and some examples of model working procedures that may help to eliminate or effectively reduce the potential hazards when working in this environment. Sections 5, 7, 10 and 11 contain specific references to pumping stations covering atmospheric hazards, operational procedures and precautions against potential hazards, emergency procedures and designing to eliminate hazards.

A3.7 MISCELLANEOUS BOOKS

Pumping station equipment and design. By L B Escritt (C R Books Ltd, 1962)

A good basic text, but a little out of date (12 chapters, 139 pages).

Pumping stations for water and sewage. By Ronald E Bartlet (Applied Science Publishers Ltd, 1974)

A fuller and more up-to-date text than the above. Covers submersible pumps and package units (12 chapters and three appendices, 150 pages).

Pumping station design. Editor-in-chief: Robert L Sanks (Butterworths, 1989)

A very detailed text covering all aspects of water, wastewater and sludge pumping station design, containing contributions from over 130 authors and contributors (29 chapters and eight appendices, 828 pages).

A3.8 MISCELLANEOUS PAPERS [UK]

Stanbridge H H: History of sewage pumping in Britain. Wat Pollut Control, 1977, p517

A brief review of sewage pumping from 1842 to 1974.

Hobson J A & Carne R J: Energy saving in sewage pumping by the use of low velocities. Wat Pollut Control, 1986, p277

The paper concludes that low-velocity pumping (about 0.4 m/s) in dual-rate systems can significantly reduce energy usage where the total has a substantial friction component. Velocities as low as 0.4 m/s did not apparently have any adverse effect on the state of the rising main, provided that higher velocities (0.75–1.0 m/s) are used periodically.

Evans G P: The optimisation of sewage pumping using micro-processors. The Public Health Engineer, 1981, p208

Paper based on WRcTR 170, see reference 10 above.

Townsend G H & Greeves I S S: The design and construction of Gallions surface water pumping station. Proc Instn Civ Engrs, Part 1, 1979, 66, p605

Large low-lift stormwater pumping station with several novel features. Pumps are located in a circular dry well and draw radially from the circular wet well surrounding it.

Lambert D J: The advantages of mutrator pipeline systems in sewage and effluent pumping. Infrastructure Renovation & Waste Control, North West Area, 8–10 April, 1986, p158

The acceptance of small-bore pipelines has meant simple and easy main-laying, often in difficult terrain, and greatly reduced pipeline costs. The correspondingly reduced retention time in the main reduces the risk of septicity. A mutrator consists of a positive displacement pump and a macerator mounted over a wet well, often at ground level.

Grimes J F, Allsop A E, Henham-Barrow J E C, Long J J & Thompson V G: Conveyance system: pumping stations and culverts. Proc Instn Civ Engrs, Civ Eng, Greater Cairo Wastewater Project, 1993, p34

Description of the design and construction of five pumping stations with large outputs and low lifts, varying from 7.2 m to 22 m.

Griffith A: Concept of buildability. Heriot-Watt University, Scotland; Reports of the Working Commissions. International Association for Bridge and Structural Engineering, Zurich, Switzerland, V 53, 1986, p105

Considers the problem of buildability – the extent to which the design of a building facilitates ease of construction – focusing upon the technological and managerial aspects of making a building more buildable following principles of design developed in house building. A case study shows how these principles react to more technologically complex design types.

A3.9 MISCELLANEOUS PAPERS [USA]

Submersible lift stations cut pump maintenance costs. Water & Sewage Works, Vol 122, No 8, 1975, p64

Conversion of most of its sewage pumping stations to the submersible type is paying off in substantial maintenance savings at the planned community of Foster City, California.

Nicolai B: Licking operational problems the hard way. Water and Wastes Engineering, Vol 14, No 9, 1977, p51

If not properly looked after, sewage pumping stations may present manifold problems, some of which can be fatal.

Ransom J E: Submersible pumps speed sewage flow. The American City and County, Vol 92, No 6, 1977, p70

Rapid growth and worn-out pumping station equipment made Florence, Kentucky, look twice at the pumps that local contractors were using to dewater construction sites.

Glidewell J D Jr: Sewage pumping built to handle future system demands. Water and Sewage Works, Vol 124, No 4, 1977, p105

A new sewage pumping station at Topeka, USA, designed for present and future demands, involves careful selection of pumps (variable-speed pumps with adequate standby facilities installed) and provision for maintenance as part of the plant operation.

Dowell J C & Wood S: Sewage pumping station split for maintenance. The American City and County, Vol 97, No 4, 1982, p60

Maintenance and economy were built in when Wichita, Kansas, completed its new sewage pumping station. A dual-flow design enables half the station to be shut down when repairs are needed.

References

1	HMSO *The Construction (Design and Management) Regulations* SI 1994:3140, London 1994
2	CIRIA *Buildability – an assessment* Special Publication 26, 1983
3	ADAMS, STEWART *Practical buildability* CIRIA Building Design Report, 1989
4	PROSSER, M J *Design of low-lift pumping stations with particular application to wastewater* CIRIA Report 121. CIRIA/BHRA fluid engineering, 1992
5	BS 8005: Part 2: 1987 *Sewerage: Guide to pumping stations and pumping mains*
6	SEWERS AND WATER MAINS COMMITTEE *A guide to sewerage operational practices* Water Services Association and Foundation for Water Research, 1991
7	HMSO *Control of Substances Hazardous to Health (COSHH)* SI 1994:3246 London, 1994
8	HMSO *Electricity at Work Regulations 1989* SI 1989:635 London, 1989
9	HMSO *Environmental Protection Act 1990* London, 1990
10	HMSO *Health and Safety at Work Act* London, 1974
11	WATER SERVICES ASSOCIATION *Sewers for adoption - a design and construction guide for developers* 4th edition, May 1995, published by WRc plc
12	PROSSER, M J *The hydraulic design of pump sumps and intakes.* CIRIA with BHRA (Fluid Engineering), 1977

13   **WATER AUTHORITIES ASSOCIATION**
Sewers for adoption – a design and construction guide for developers
3rd edition, published by Water Research Centre (1989) plc